I0787959

Graphene, Ge/III-V, Nanowires, and Emerging Materials for Post-CMOS Applications 4

Editors:

Y. Obeng
National Institute of Standards and Technology
Gaithersburg, Maryland, USA

S. De-Gendt
IMEC
Leuven, Belgium

Z. Karim
AIXTRON, Inc.
Sunnyvale, California, USA

D. Misra
New Jersey Institute of Technology
Newark, New Jersey, USA

P. Srinivasan
Texas Instruments
Dallas, Texas, USA

Sponsoring Divisions:

 Dielectric Science & Technology

 Electronics and Photonics

Published by
The Electrochemical Society

65 South Main Street, Building D
Pennington, NJ 08534-2839, USA

tel   609 737 1902
fax 609 737 2743
www.electrochem.org

ecstransactions ™

Vol. 45, No. 4

Copyright 2012 by The Electrochemical Society.
All rights reserved.

This book has been registered with Copyright Clearance Center.
For further information, please contact the Copyright Clearance Center,
Salem, Massachusetts.

Published by:

The Electrochemical Society
65 South Main Street
Pennington, New Jersey 08534-2839, USA

Telephone 609.737.1902
Fax 609.737.2743
e-mail: ecs@electrochem.org
Web: www.electrochem.org

ISSN 1938-6737 (online)
ISSN 1938-5862 (print)
ISSN 2151-2051 (cd-rom)

ISBN 978-1-56677-956-2 (Hardcover)
ISBN 978-1-60768-314-8 (PDF)

Printed in the United States of America.

PREFACE

The *Fourth International Symposium on Graphene, Ge/III-V and Emerging Materials For Post-CMOS Applications* was held from May 6-10, 2012, at the Washington State Convention Center and the Sheraton Seattle in Seattle, WA, at the 221st Meeting of the Electrochemical Society as symposium E2, organized by the Dielectric Science and Technology Division.

The first three symposia in this series drew tremendous response from people all over the world. The objective of this symposium is to assemble researchers and technical personnel from Industry, Universities and Government Laboratories around the globe to address current and future issues of fundamental material science, characterization and applications of emerging materials designed for alternative information processing technologies to replace CMOS.

As industry moves into the functional diversification and equivalent scaling paradigms, materials and device needs require exploration far beyond the boundaries of conventional semiconductor technologies and applications. With these shifts in the landscape, the focus is now on creating high value micro/nanoelectronics systems, to drive new technological possibilities and unlimited application potential. While grapheme-based electronics afford strong potential candidates for building ultra-low power electronics, Ge-III/V devices are attractive options for developing high-speed applications. Both the areas are of technical significance and importance to this symposium. Whereas this symposium particularly emphasizes "Beyond CMOS" materials, integration schemes and the technology development associated with such integration, the information shared and the contents of this volume are also pertinent to "More-Than-Moore" world.

Compared to previous years, a considerable amount of Graphene and III-V development has taken place. The fundamental material properties are better understood, the knowledge on growth kinetics and transport has improved, solutions towards characterization have been identified and finally the art of integrating this material into the process has taken another step.

The organizers would also like to express special thanks to the speakers for their interest in and support of the symposium, by submitting high-quality abstracts and preparing their manuscripts at short notice. We are also grateful to the staff of The Electrochemical Society, particularly John Lewis, Paul Urso, and Beth Anne Stuebe, who helped us at various stages preceding, during, and after the symposium, and during the publication period. Finally, the success of the symposium would not have been possible without the financial support given by the sponsoring ECS Divisions (Dielectric Science and Technology, and New Technology Committee) and by our industry sponsor—AIXTRON. Their sustained interest and sponsorship are highly valued.

We hope to continue publishing the proceedings of future editions of this symposium and collecting them into an *ECS Transactions* series which will benefit every individual in this field, as a useful and valuable reference. Once again, we thank the speakers for their contributions to their exciting symposium at Seattle, Washington.

Yaw Obeng
Purushothaman Srinivasan
Zia Karim
Stefan De Gendt
Durga Misra

May 2012

***ECS Transactions*, Volume 45, Issue 4**
Graphene, Ge/III-V, Nanowires, and Emerging Materials for Post-CMOS Applications 4

Table of Contents

Preface   *iii*

Chapter 1
Physics and Technology of Graphene

Bilayer Pseudo-Spin Field Effect Transistor (BiSFET): Concepts and Critical Issues   3
for Realization
     L. F. Register, X. Mou, D. Reddy, W. Jung, I. Sodemann, D. Pesin, A. Hassibi,
     A. H. MacDonald, and S. K. Banerjee

Current Switching in Crossed Graphene Nanoribbons   15
     R. K. Lake and K. Habib

Chapter 2
Performance and Metrology of Graphene

Device Characteristics of In Situ CCVD Grown Bilayer Graphene FETs at Elevated   23
Temperatures
     P. Wessely, F. Wessely, E. Birinci, B. Riedinger, and U. Schwalke

Surface Potential Measurements of Reconfigurable p-n Junctions in Graphene   31
     Y. Wang and R. E. Geer

In Situ Electrical Studies of Ozone Based Atomic Layer Deposition on Graphene   39
     S. Jandhyala, G. Mordi, B. Lee, and J. Kim

Chapter 3
Graphene Growth and Characterization

Direct Graphene Growth on Oxides: Interfacial Interactions and Band Gap Formation   49
     J. A. Kelber, M. Zhou, S. Gaddam, F. L. Pasquale, L. Kong, and P. A. Dowben

Aberration Corrected Microscopy of CVD Graphene and Spectroscopic Ellipsometry   63
of Epitaxial Graphene and CVD Graphene for Comparison of the Dielectric Function
  F. Nelson, D. Prasad Sinha, E. Comfort, J. Lee, A. Diebold, A. Sandin,
  D. B. Dougherty, and J. E. Rowe

Modeling the Growth of SWNTs and Graphene on the Atomic Scale   73
  E. C. Neyts and A. Bogaerts

Large Area Mapping of Graphene Grain Structure and Orientation   79
  H. Floresca, D. Hinojos, N. Lu, J. Chan, L. Colombo, R. Wallace, J. Wang, J. Kim,
  and M. J. Kim

Reduced Pressure-Chemical Vapor Deposition of High Quality Ge Layers on SiGe/Si   83
Superlayers for Microelectronics and Optoelectronics Purposes
  D. Chen, Z. Xue, S. Liu, and M. Zhang

Chapter 4
Ge and III/V Technologies

III-Sb MOSFETS : Opportunities and Challenges   91
  A. Nainani, Z. Yuan, A. Kumar, B. R. Bennett, J. Boos, and K. C. Saraswat

Passivation Challenges with Ge and III/V Devices   97
  S. Sioncke, D. Lin, L. Nyns, A. Delabie, A. Thean, N. Horiguchi, H. Struyf,
  S. De-Gendt, and M. Caymax

Investigations on Thermal Stress Relief Mechanism Using Air-Gapped SiO_2 111
Nanotemplates during Epitaxial Growth of Ge on Si and Corresponding Hole Mobility
Improvement
  S. Ghosh, D. Leonhardt, and S. M. Han

Integration of InGaAs Channel n-MOS Devices on 200mm Si Wafers Using the   115
Aspect-Ratio-Trapping Technique
  N. Waldron, G. Wang, N. D. Nguyen, T. Orzali, C. Merckling, G. Brammertz,
  P. Ong, G. Winderickx, G. Hellings, G. Eneman, M. Caymax, M. Meuris,
  N. Horiguchi, and A. Thean

Deterministic Assembly of $In_{0.53}Ga_{0.47}As$ p^+-i-n^+ Nanowire Junctions for Tunnel Transistors   129
  M. Kuo, J. Li, H. Liu, A. Vallett, D. Mohata, S. Datta, and T. S. Mayer

Desorption of Ge Species during Thermal Oxidation of Ge and Annealing of   137
HfO_2/GeO_2 Stacks
  C. Radtke, G. Rolim, S. da Silva, G. Soares, C. Krug, and I. Baumvol

Chapter 5
III-V Integration

New Method to Produce High-Quality Epitaxial Ge on Si Using SiO_2-Lined Etch Pits and Epitaxial Lateral Overgrowth for III-V Integration   147
D. Leonhardt and S. M. Han

VO_2, a Metal-Insulator Transition Material for Nanoelectronic Applications   151
K. M. Martens, I. P. Radu, G. Rampelberg, J. Verbruggen, S. Cosemans, S. Mertens, X. Shi, M. Schaekers, C. Huyghebaert, S. De-Gendt, C. Detavernier, M. Heyns, and J. A. Kittl

Demonstration of Single Crystal GaAs Layers on CTE-Matched Substrates by the Smart Cut Technology   159
T. Jouanneau, Y. Bogumilowicz, P. Gergaud, V. Delaye, J. Barnes, V. Klinger, F. Dimroth, A. Tauzin, B. Ghyselen, and V. Carron

Characterization of Rapid Melt Growth (RMG) Process for High Quality Thin Film Germanium on Insulator   169
N. Zainal, S. Mitchell, D. W. McNeill, M. F. Bain, B. Armstrong, P. T. Baine, D. Adley, and T. S. Perova

Chapter 6
Fabrication and Characterization of III-V's

In Situ As_2 Decapping and Atomic Layer Deposition of Al_2O_3 on n-InGaAs(100)   183
J. Ahn, B. Shin, and P. McIntyre

Germanium Doping, Contacts, and Thin-Body Structures   189
R. Duffy and M. Shayesteh

Germanium on Insulator (GOI) Structure Using Hetero-Epitaxial Lateral Overgrowth on Silicon   203
J. Nam, T. Fuse, Y. Nishi, and K. C. Saraswat

Multiple-Gate $In_{0.53}Ga_{0.47}As$ Channel n-MOSFETs with Self-Aligned Ni-InGaAs Contacts   209
X. Zhang, H. Guo, X. Gong, C. Guo, and Y. Yeo

Sub-100nm Non-Planar 3D InGaAs MOSFETs: Fabrication and Characterization   217
J. J. Gu and P. D. Ye

vii

Non-Destructive, Large-Scale Imaging of Anti-Phase Disorder in GaP Epilayers on   231
Si(001) Using Low-Energy Electron Microscopy
    B. Borkenhagen, H. Döscher, T. Hannappel, G. Lilienkamp, and W. Daum

Author Index   241

Facts about ECS

The Electrochemical Society (ECS) is an international, nonprofit, scientific, educational organization founded for the advancement of the theory and practice of electrochemistry, electrothermics, electronics, and allied subjects. The Society was founded in Philadelphia in 1902 and incorporated in 1930. There are currently over 7,000 scientists and engineers from more than 70 countries who hold individual membership; the Society is also supported by more than 100 corporations through Corporate Memberships.

The technical activities of the Society are carried on by Divisions. Sections of the Society have been organized in a number of cities and regions. Major international meetings of the Society are held in the spring and fall of each year. At these meetings, the Divisions and Groups hold general sessions and sponsor symposia on specialized subjects.

The Society has an active publications program that includes the following.

Journal of The Electrochemical Society — JES is the peer-reviewed leader in the field of electrochemical and solid-state science and technology. Articles are posted online as soon as they become available for publication. This archival journal is also available in a paper edition, published monthly following electronic publication.

Electrochemical and Solid-State Letters — ESL is the first and only rapid-publication electronic journal covering the same technical areas as JES. Articles are posted online as soon as they become available for publication. This peer-reviewed, archival journal is also available in a paper edition, published monthly following electronic publication. It is a joint publication of ECS and the IEEE Electron Devices Society.

Interface — *Interface* is ECS's quarterly news magazine. It provides a forum for the lively exchange of ideas and news among members of ECS and the international scientific community at large. Published online (with free access to all) and in paper, issues highlight special features on the state of electrochemical and solid-state science and technology. The paper edition is automatically sent to all ECS members.

Meeting Abstracts (formerly Extended Abstracts) — Abstracts of the technical papers presented at the spring and fall meetings of the Society are published on CD-ROM.

ECS Transactions — This online database provides access to full-text articles presented at ECS and ECS-sponsored meetings. Content is available through individual articles, or as collections of articles representing entire symposia.

Monograph Volumes — The Society sponsors the publication of hardbound monograph volumes, which provide authoritative accounts of specific topics in electrochemistry, solid-state science, and related disciplines.

For more information on these and other Society activities, visit the ECS website:

www.electrochem.org

CHAPTER 1

PHYSICS AND TECHNOLOGY OF GRAPHENE

Bilayer Pseudo-Spin Field Effect Transistor (BiSFET):
Concepts and Critical Issues for Realization

L. F. Register[a], X. Mou[a], D. Reddy[a], W. Jung[a], I. Sodeman[b], D. Pesin[b],
A. Hassibi[b], A. H. MacDonald[b] and S. K. Banerjee[a].

[a] Department of Electrical and Computer Engineering and Microelectronics Research
Center, The University of Texas at Austin, Austin, Texas 78712
[b] Department of Physics, The University of Texas at Austin, Austin, Texas 78712

> The Bilayer pseudo-spin Field Effect Transistor (BiSET) has been
> proposed as one means of taking advantage of possible room
> temperature superfluidity in two graphene layers separated by a
> thin dielectric. In principle, the switching energy per device could
> be on the scale of 10 zJ, over two orders of magnitude below
> estimates for "end-of the roadmap" CMOS transistors. However,
> achieving both the goal of room temperature superfluidity and
> harnessing it for low-power switching pose substantial challenges,
> both theoretical and experimental. In this work we review the
> basic graphene superfluidity and BiSFET concepts, our current
> understanding—and limits to that understanding—of the
> requirements for condensate formation, and how these
> requirements could impact BiSFET design.

Introduction

In 2008 Min, Bistritzer, Su and MacDonald pointed to the possibility of room
temperature superfluid condensation in two graphene layers separated by a thin dielectric
[1]. Although it has not yet been experimentally realized, there has been debate about
this possibility, and the best estimates of the required conditions have been refined, recent
theory continues to point in this direction [2].

If such room temperature condensation can be achieved and achieved practically, no
doubt many applications will result. The Bilayer pseudo-Spin Field Effect Transistor
(BiSET) has been proposed as one means of taking advantage of such room temperature
superfluidity for "beyond CMOS" logic applications [3]. In principle, the switching
energy per device could be on the scale of 10 zJ (10×10^{-18} J), more than two orders of
magnitude below estimates for "end of the roadmap" CMOS transistors [4]. However,
particularly with our improving understanding of the requirement for superfluity in this
double graphene layer system, it's clear that achieving both the goal of room temperature
superfluidity and harnessing it for low-power switching pose substantial challenges.

In this work we review the basic principles behind superfluidity in graphene and the
BiSFET device concept, and discuss how our improved understanding of the
requirements for superfluidity could impact design of the proposed BiSFET, or, in the
short term, test structures for experimentally observing the superfluid condensate.

Conceptual Basis for Room Temperature Superfluidity in Paired Graphene Layers

Although it has not yet been experimentally realized in paired graphene layers, the signature of such coherent interlayer superfluid condensation has been seen at low temperatures in high magnetic fields (the latter to form effective electron-hole systems within Landau levels) in III-V double quantum well systems [5,6]. In the paired graphene layers considered here, however, high and potentially room temperature condensation is favored by a synergy of graphene properties: the ability to use closely spaced atomically thin layers to maximize the interlayer Coulomb interaction; symmetric electron and hole band structures over the energy ranges of interest that allow accurate nesting between the electron and hole two-dimensional Fermi surfaces; a zero bandgap which allows all of any interlayer electrostatic potential difference to be used to induce electrons and holes; and a low density of states that leads to the required high Fermi energies relative to the Dirac points at relatively low carrier densities.

The basic concept of interlayer condensation is illustrated in Fig. 1 for two graphene layers with a large interlayer potential energy (not voltage) difference, which should be approximately $16k_BT$ (Boltzmann constant times temperature) or greater [1], corresponding to electron (n) and hole (p) concentrations of $4.5\times10^{12}/cm^2$ or greater at room temperature. If no coupling is assumed between layers, the wave-functions are localized to one layer or the other, and one simply obtains two offset monolayer graphene energy band structures with sizable electron and hole populations. However, in principle, interlayer coherence in the vicinity of the now anti-crossing between the bandstructures of the two layers and a many-body non-local exchange potential between the layers can self-consistently support each other. Within a Fock mean-Field approximation and neglecting free-carrier screening, the non-local exchange potential (i.e., containing off-diagonal potential elements within a Hamiltonian matrix) is

$$V_{\text{Fock,int}}(\mathbf{r}_T,\mathbf{r}_B) \approx -\frac{e^2}{4\pi\varepsilon\sqrt{|\mathbf{r}_T-\mathbf{r}_B|^2+d^2}}\sum_{\mathbf{k}} f_{\mathbf{k},m}\psi_{\mathbf{k},m}(\mathbf{r}_T,T)\psi_{\mathbf{k},m}^*(\mathbf{r}_B,B) \qquad (1)$$

where ε is assumed uniform here for the sake of clarity, d is the separation between the graphene layers, $\mathbf{r}$ indicates position along the plane of the layers, T and B indicate which layer, $|\mathbf{r}_T-\mathbf{r}_B|$ is the magnitude of the separation within the plane of the layers, f is the occupation probability, and the ψ are the individual quasi-single-particle wave-functions labeled by the two-dimensional wave-vector $\mathbf{k}$ and band m. At least some of these wave-functions must be delocalized between the layers for a nonzero interlayer exchange interaction. This exchange interaction is maximized when the Fermi level falls in the middle of the anti-crossing band gap. And with the Fermi-level in the middle of the band gap, this latter interlayer-coherent many-body state is the overall energetically favorable one as the energies of the occupied states are reduced via the anti-crossing. This coupling also can be interpreted as coherent exciton condensation between the electrons in bottom layer and holes in the top layer.

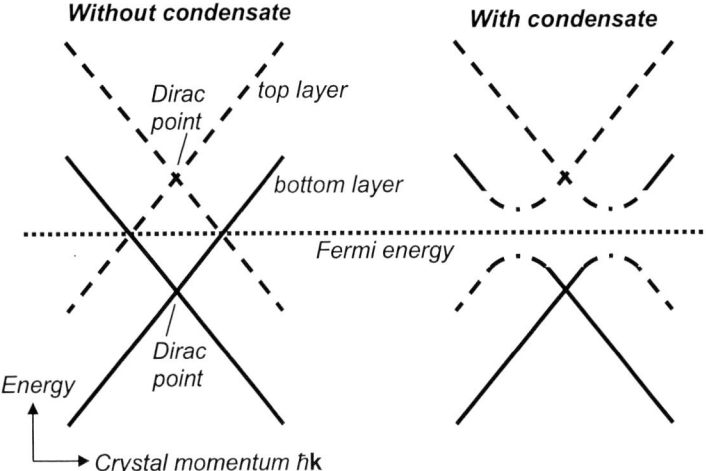

Figure 1. Band structure of two graphene layers with potential offset, without significant interlayer exchange (or other) coupling (left), and with significant exchange coupling (right). In the latter case, in the vicinity of the anti-crossing, the quasi-single-particle states are delocalized between the two layers.

The BiSFET Device and Circuit Concept

The BiSFET as originally proposed is illustrated in Fig. 2 along with its qualitatively anticipated current-voltage characteristics exhibiting two critical aspects of the proposed BiSFET, it's potential extremely low voltage operation and very non-MOSFET-like output characteristics. The V_{Gn} and V_{Gp} gates in this layout, perhaps as short as 10 nm, were intended for two purposes: The first was to create large and nominally equal and opposite charge densities to allow formation of the superfluid between the gates, although through work-function engineering and or back-gating rather than application of large voltages on both gates. The enhanced interlayer coupling associated with the superfluid should then allow greatly enhanced interlayer tunneling (see arrows in Fig. 2) up to some critical current density dependent on the overlap of the bare coupling and the quasi-particle wave-functions within the condensed superfluid state [3,7]. After reaching this critical current, the superfluid and associated enhanced current is suspected to collapse with increasing interlayer voltage magnitude $|V_n - V_p|$, much as for a Josephson junction, at a critical voltage potentially less than $k_B T$ and perhaps only a few meV. The second function of these gates in the initial design was to allow for imbalancing the charge distributions through application of quite small gate voltages, perhaps of the scale of room temperature $k_B T$, to weaken the superfluid (see below), and, thus, reduce the critical current density and associated critical voltage.

While, as to be discussed later, the physical layout will likely have to be changed, the device current-voltage characteristics will likely remain at least qualitatively the same. Determining if and how one might use such devices to achieve basic logic functions as well as more complicated circuits was one of the first tasks. We developed a simple

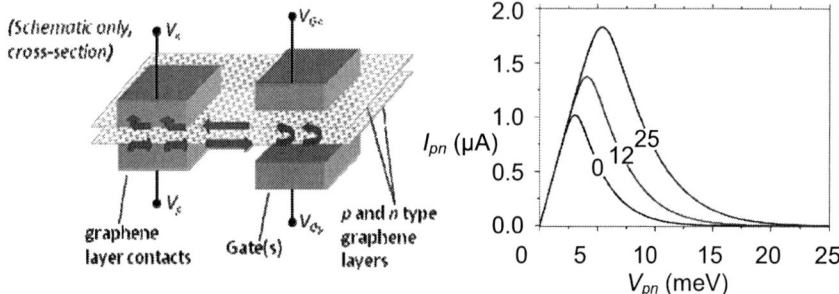

Figure 2. Schematic diagram of the BiSFET as originally proposed (left), and qualitatively expected current voltage characteristics (right) for initially balanced charge densities. The curves in the latter figure are labeled by the input voltage in mV to one gate that unbalances the charge.

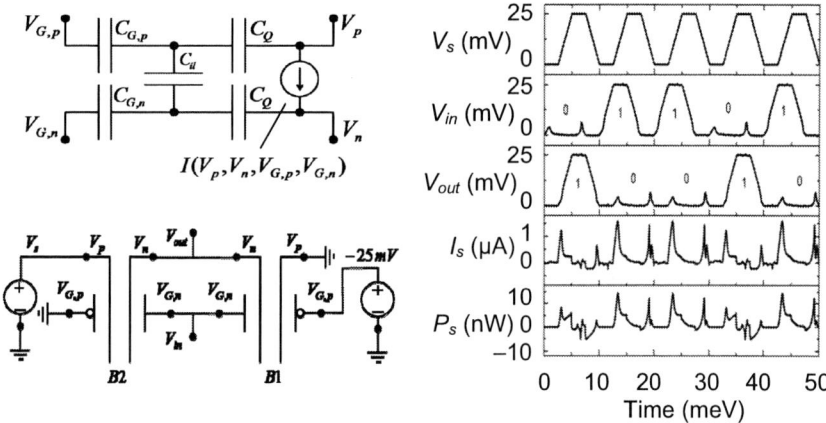

Figure 3. SPICE circuit model of the BiSFET (top left), A two-BiSFET inverter with BiSFET B2 charge-balanced and BiSFET B1 charge-unbalanced at $V_{in} = 0$ V, and vice versa with $V_{in} = 25$ mV (Bottom left), and SPICE-simulated circuit characteristics (right). The integrated switching energy consumption per BiSFET was approximately 10 zepto-Joules with a four-inverter load

SPICE model of the BiSFET as per Fig. 3, and, with the aid of a four-phase clock low-voltage (0 to 25 mV) power supply, we were able to demonstrate a full range of basic logic gates [3,8] and more complex elements such as an four-bit adder recently [9]. For the illustrative inverter circuit also of Fig. 3, as the power supply voltage and the corresponding current through the inverter is increased, the critical current density of the charge-unbalanced BiSFET, as determined by the input voltage V_{in}, is reached forcing it into the a high-resistance "off" state and reducing the current again, while the other BiSFET remains in its low resistance "on" state. With the parameters used in those simulations and without counting the clock, the integrated switching energy consumption

per BiSFET was approximately 10 zepto-Joules with a four-inverter load. This value is well over two orders of magnitude below end of the roadmap CMOS values [4]. Other logic gates, using more devices, work in a similar fashion with similar power consumption per device. The primary reason for such a drastic reduction in power consumption is simply the dependence of power on the square of these very small voltages. We have since found a way to generate the 25 mV four-phase clocks from CMOS circuitry with perhaps greater than 70% efficiency (which also may be useful for other low-voltage device concepts) [10]. However, each inverter switches each clock cycle so that the activity factor is effectively unity, so some of that advantage is lost. On the other hand, the circuits should work quite nicely in a pipelined architecture. Of course, these numbers should only be considered very rough estimates with the uncertainties involved. However, even remotely approaching such switching energies would still represent a substantial breakthrough.
.

Challenges to Realization

Even if theory holds, there are a number of challenges to achieving room temperature superfluidity including choosing the appropriate dielectric materials, minimizing the spacing between dielectric layers while still limiting the bare/single-electron coupling between the layers, and understanding the effects of charge imbalance and rotation between the graphene layers.

Charge Imbalance

As discussed above, the strength of the exchange interaction is maximized and the energy of the condensed system for a given exchange interaction is minimized when the Fermi level lies at the middle of the anti-crossing band gap. The reason is that the quasi-single-particle states which contribute most strongly to the exchange interaction are those occupied states near the band gap. However, an occupied state of the value $\mathbf{k}$ above the anti-crossing band gap contributes almost precisely (precisely in the single π-orbital tight-binding model of graphene) opposite that of an occupied state of the same $\mathbf{k}$ below the gap. For this reason, charge imbalance, which leads to either emptying of states below the gap or occupation of states above the gap and a self-consistent shift of the Fermi level toward the bottom or top edge of the gap, respectively, was the originally proposed mechanism of gating the BiSFETs. It now appears that such a gating approach will not work for other reasons to be discussed (but we will also discuss an alternative). However, much like any device sensitivity, understanding how much charge imbalance between layers is acceptable for achieving superfluidity remains important to BiSFET device and circuit design, or simply experimental observation of the superfluid. For example, long range disorder, which would produce so-called charge pudding for lower carrier densities, could still produce charge imbalance at these high carrier densities.

The effect of such charge imbalance on the condensate's strength as measured by its anti-crossing band gap calculated via the unscreened Foch exchange interaction of Eq. (1) are illustrated in Fig. 4 [11]. Precise effects varied with the exact structure and temperature considered. However, as per this example, overall we found that a roughly 10% charge imbalance, $(n-p)/(n+p)$, had limited effects on the condensate, and that the condensate survived to some degree up to roughly a 20% charge imbalance.

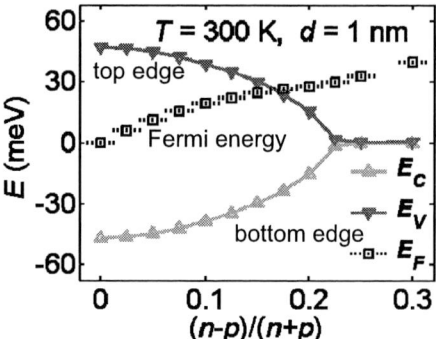

Figure 4. Anti-crossing energy band gap edges and Fermi level as a function of carrier imbalance between top layer electron density and bottom layer hole density (increasing electron and decreasing hole concentrations, specifically here, but the effect is symmetric). Here, the graphene layers are separated by 1 nm, there is uniform overall relative dielectric constant ε_r of 3 and free-carrier screening, there is an interlayer potential (not voltage) difference of 0.5 eV corresponding to nominal electron and hole populations of somewhat over $5\times10^{12}/cm^2$ each, and the temperature is 300 K.

Interlayer Rotation

Formation of the condensate requires that the bare coupling between the graphene layers be limited, such as by an interlayer dielectric. However, while one may speculate about, e.g., epitaxially grown graphene/hexagonal-boron-nitride/graphene stacks, currently there is no practical means to align graphene layers separated by a dielectric. Eventually, it appears likely that such control to produce aligned or even controllably misaligned graphene layers will be necessary for the purpose of creating BiSFETs. However, as the bare coupling between misaligned graphene layers is known to be quite sensitive to and strongly diminished by interlayer rotation or twisting [12], it may even be possible to employ interlayer rotation in lieu of an interlayer dielectric.

But what of interlayer rotation and the condensate itself? As to be expected via Eq. (1) with the interlayer atomic separation large compared to intralayer separation, relative lattice translations or offsets have essentially no effect on the condensate strength nor even the two-dimensional shape of the non-local exchange potential as a function of r_T and r_B [11], at least for the "spontaneous condensate" calculated in the absence of bare tunneling (although significant bare coupling can alter this result [7]). Less obviously, we have also found recently that rotation has essentially no effect on the strength of this spontaneous condensate, despite significantly altering the two dimensional shape of the non-local exchange potential [13], as illustrated in Fig. 5. In essence, the condensate is quite adaptable and adjusts its shape as needed to maximize is strength. As a result, however, while control of the interlayer rotation may not be possible in the short term, neither may it be necessary in the short term for efforts to simply observe the condensate.

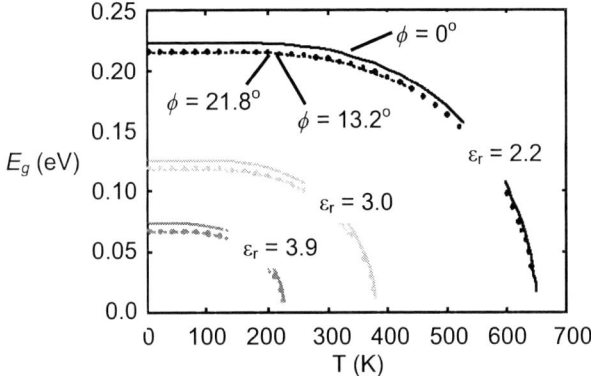

Figure 5. Effects of commensurate interlayer rotation angle ϕ (solid line = 0°, dotted line = 21.8° and dashed line = 13.2°) for varying relative dielectric constants $\underline{\varepsilon_r}$ as a function of temperature in Kelvin, as obtained via Eq. (1) without free carrier screening. Only a small and essentially angle-independent change in the 0 K band gap is observed for non-zero interlayer angles, and essentially no change in the critical temperature at which the condensate collapses is observed with interlayer rotation.

Free-Carrier and Dielectric Screening

Screening is probably the most important issue for observing condensate formation experimentally and, more challenging, using it for practical applications including the BiSFET. For example, the exchange interaction, which can extend over a few nm in the plane of the graphene layers, decays with reciprocal of the dielectric constant ε_r for all values of $\mathbf{r_T} - \mathbf{r_B}$ within the approximation of Eq. (1), while the exchange interaction only decays as one of the interlayer distance d for $\mathbf{r_T} - \mathbf{r_B} = 0$. Thus, we want overall low-k dielectrics in contrast to the needs for MOSFET gate dielectrics. Furthermore, again in contrast to the case for MOSFET gate dielectrics, the field lines extend beyond the interlayer dielectric for these carrier-carrier interactions. Indeed, the dielectrics above and below the graphene are probably more important than the interlayer dielectric. These effects of differing dielectric stacks can be calculated in a straightforward matter given the dielectric constants, if not always with straightforward results. For example, with the charges on the opposing interfaces between the dialectics, putting a lower-k dielectric between two higher-k dielectrics is actually counterproductive.

Choosing the appropriate dielectric constants is not as straightforward, however. If we were performing gate capacitance calculations we would clearly want to use the static (low-frequency) dielectric constant. However, the carriers in graphene travel approximately 100 nm in the classical oscillation period of a 40 meV optical phonon, where the polar optical phonon frequency roughly marks the transition frequency between low- and high-frequency dielectric constants in polar materials. In contrast, the range of the exchange potential is only on the scale of perhaps 5 nm. Therefore, from at least a classical perspective, it would seem unlikely that the crystal ions would be able to respond quickly enough to help screen the coulomb-mediated exchange interaction in this system. Although a more rigorous quantum analysis has yet to be performed, this argument suggests that perhaps the high-frequency dielectric constant should be used. (We note that these same arguments do not apply to quasi-static capacitance calculations even though the surface charge is composed of many fast moving individual carriers.)

More controversial have been the effects of free-carrier screening. However, recent calculations addressing dynamic free-carrier screening with a free-carrier population self-consistently dependent on condensate continue to point toward the possibility of room temperature condensation [2], as will also be discussed elsewhere within these proceedings. However, even if still possible, free-carrier screening clearly makes the task more difficult. In particular, it's likely that free carrier screening will have to be compensated for by use of even lower-k dielectric materials than previously expected, as illustrated by preliminary results from that effort shown in Fig. 6. (Final results vary slightly quantitatively, but likely not beyond the uncertainty of the results). As illustrated in the figure, the required value of the dimensionless coupling constant $e^2/(4\pi\varepsilon\hbar v_F)$ required to achieve a condensate within the approximate range of interest for the BiSFET increases by a factor in the rough vicinity of 2.4. Here, v_F is the carrier Fermi velocity. Noting that the free-space value of the above coupling constant is approximately 2.2 and that this preliminary estimate of the required coupling constant is approximately 1.4 in the range of interest, the effective relative dielectric constant needs to be quite small indeed even for high-k dielectrics, around 1.6 based on these numbers.

To address the net effect of all of these dielectric and free-carrier screening considerations, we considered a variety of dielectric stacks as illustrated in Fig. 7. Note

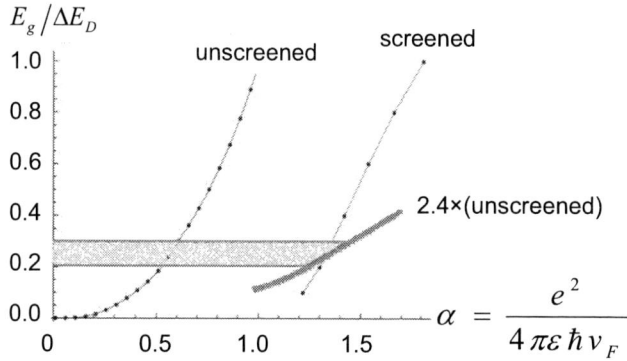

Figure 6. Estimated minimum value of the dimensional coupling constant $e^2/(4\pi\varepsilon\hbar v_F)$ required for condensation as a function of one-half the condensate anti-crossing band gap E_g normalized by the Fermi energy E_F, for no free-carrier screening and self-consistent dynamic screening within a mean-field approximation [2]. For room temperature condensation, the magnitude of the interlayer potential ΔE_D difference should be approximately 450 meV or greater as per prior discussion. The region between the horizontal lines is the approximate range of interest; the short tilted line corresponds to multiplying the unscreened result by 2.4. In other words, roughly speaking, the effect of free-carrier screening is analogous to increasing the dielectric constants by a factor of 2.4 in the absence of free-carrier screening within this region.

that absent free-carrier screening, any one of these systems could have shown some degree of condensation for some temperatures for clean samples in principle. However, consistent with measurements to date, these results with free-carrier screening suggest that only the more exotic systems perhaps including partial to complete vacuums in the

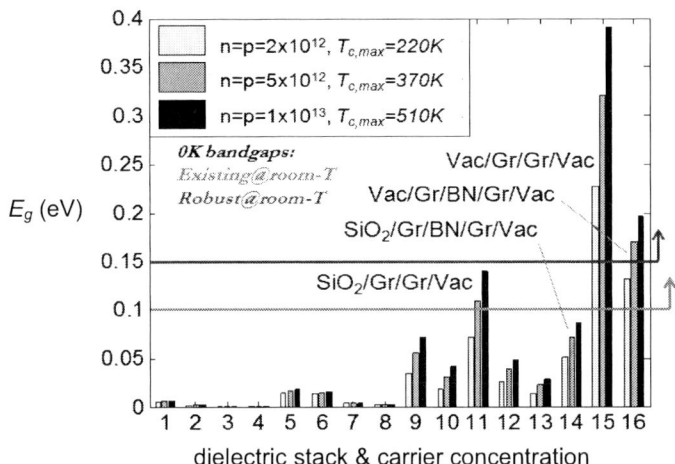

Figure 7. Estimated 0 K condensate anti-crossing band-gaps for the following graphene-dielectric stacks, assuming high-frequency dielectric constants and representing free carrier screening via a factor of 2.4 increase in modeled dielectric values over their experimental values to represent free-carrier screening, as discussed in the text: (1) SiO2 substrate, Al_2O_3* interlayer dielectric, Al_2O_3* top dielectric, d = 4nm, (2) SiO$_2$ substrate, Al_2O_3** interlayer dielectric, Al_2O_3** top dielectric, d = 4nm, (3) SiO$_2$ substrate, Al_2O_3* interlayer dielectric, Al_2O_3* top dielectric, d = 8 nm, (4) SiO$_2$ substrate, Al_2O_3** interlayer dielectric, Al_2O_3** top dielectric, d=8 nm (5) SiO$_2$ substrate, Al2O$_3$* interlayer dielectric, vacuum top dielectric, d = 4 nm (6) SiO$_2$ substrate, Al_2O_3** interlayer dielectric, vacuum top dielectric, d = 4 nm (7) SiO$_2$ substrate, Al_2O_3* interlayer dielectric, vacuum top dielectric, d = 8 nm, (8) SiO$_2$ substrate, Al_2O_3** interlayer dielectric, vacuum top dielectric, d = 8 nm, (9) SiO$_2$ substrate, touching (rotated) Gr bilayer, Al_2O_3* top dielectric; (10) SiO$_2$ substrate, touching Gr/Gr bilayer, Al_2O_3** top dielectric; (11)SiO$_2$ substrate, touching Gr/Gr bilayer, vacuum top dielectric; (12) SiO$_2$ substrate, BN monolayer interlayer dielectric Al_2O_3* top dielectric; (13) SiO$_2$ substrate, BN monolayer interlayer dielecric, Al_2O_3** top dielectric; (14) SiO$_2$ substrate, BN interlayer dielectric, vacuum top dielectric; (15) Suspended touching Gr/Gr bilayer; and (16) Suspended Gr/BN/Gr trilayer. The touching graphene layers are assumed to be rotated. (*Thin layer measure value on graphene, possibly representing some interface effects; **Bulk value). The two horizontal lines, from bottom up, correspond to the rough point of condensate formation, and that of temperature independent condensates, respectively. Three different electron and hole carrier concentrations are considered, but only the higher two are expected to be able to produce a room temperature condensate because E_F is insufficient for the lower concentrations as previously discussed. The lower was included because of its relevance for lower-temperature measurements. These dielectric requirements and, thus, the horizontal line positions are approximate, however.

vicinity of the condensate may be sufficient to produce a measureable condensate. Of course, the values of Fig. 6 are still approximate, and those of Fig. 7 are even more so.

The BiSFET and Condensate Test Structures Reconsidered

The increased understanding of the requirements for condensate realization and control has implications for BiSFET design and, more immediately, for experimental observation of the condensate.

As previously noted, it had been proposed originally that switching could be achieved via small voltages shifts applied to perhaps just one of the BiSFET gates used to create the condensate, to induce charge imbalance and, thus, reduce the critical interlayer current density. Instead of applied voltages, work-function engineering would be used for that gate to initially establish the condensate. However, there is a need for vacuums and/or other very low-k dielectrics near the region of condensation, including above and below the graphene layers. Having nearby metallic gates would also provide further screening and, thus, is to be avoided. And, although the condensate is more sensitive to the dielectrics than to the interlayer spacing, minimizing the interlayer spacing remains important. Moreover, the calculated robustness of the condensate to charge imbalance was perhaps even stronger than originally expected. The potential robustness to charge imbalance and the needs for low-k gate dielectrics, large gate-to-graphene spacing and small interlayer graphene spacing, however, mean that use of work-function engineering even for one of the condensate-creating gates will be insufficient to create the required carrier concentrations for room temperature operation. Instead, large DC voltages will be required. However, the latter requirement will preclude the gates that are used to create the condensate from also being used for the envisioned low-voltage switching.

Alternatively, rather than varying the critical current density, it should be possible to use one or more gates lying before or after the region of condensation, or both, and only partially across the path of current flow to control the local current density to provide switching, as schematically illustrated in Fig. 8. In this scenario, rather than reducing the critical interlayer current density, the control gate(s) used to induce switching would increase the local interlayer current density in a portion of the region of condensation, leading to a local breakdown of the condensate which would then, due to lack of interlayer current flow where the condensate has broken down, spread to the entire region of condensation. Moreover, the voltages on the gates used to create the condensate could be readily optimized after fabrication. The latter voltages would no doubt be large, but they need never be switched, only turned on and off when the circuit is powered up and down. And, with the gates further away from the graphene layers, the now-parasitic capacitive coupling to these latter gates would be minimized. And high-k gate dielectrics could now be used for the control gate.

Furthermore, as suggested previously, with the robustness of the condensate to interlayer graphene rotation, it might be possible to use controlled rotation in lieu of an interlayer dielectric to define the bare coupling which, in turn, defines the critical current. Previously, the strong capacitive coupling between such layers would, again, have been problematic for creating and switching the condensates with the same gates. However,

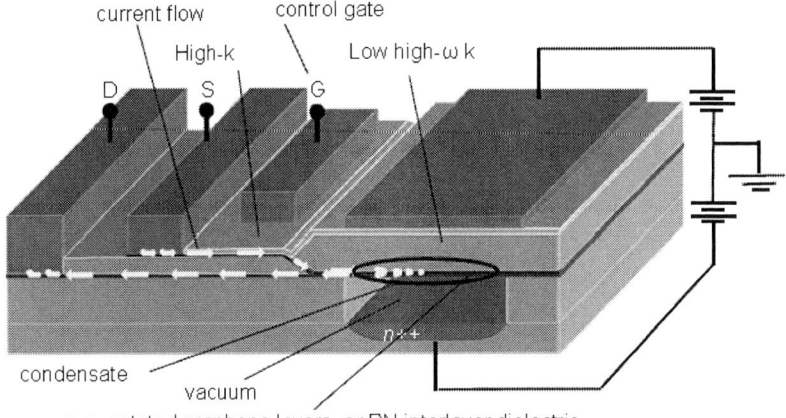

Figure 8. Schematic illustration of an alternative BiSFET design using separate gates for control and creation of the condensate, the former via a small time-dependent input voltage signal and the latter via fixed large DC voltages.

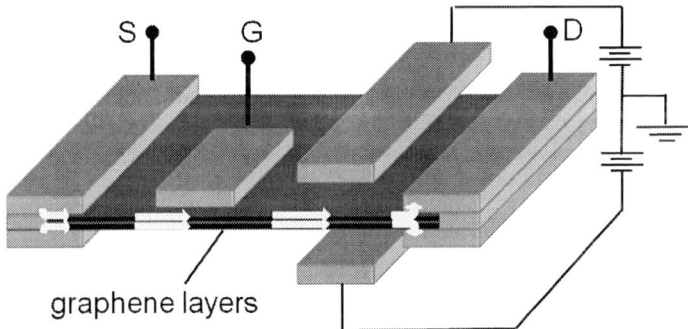

Figure 9. (Still more) schematic illustration of a structure for searching for the experimental signature of the condensate, using transport along end-shorted graphene layers, perhaps rotated or separated by a thin dielectric. Here, the gate is optional.

with the functions of the control/switching gate(s) and those used to create the condensate separated, this problem is somewhat alleviated.

The requirements for simply observing the condensate are more limited than those for creating BiSFETs. For example, as diagrammatically illustrated in Fig. 9, it may be possible to observe the condensate with in-plane (vs. interlayer) currents that would not require separate contacts to the two graphene layers nor depend on the details of the bare interlayer coupling beyond it being limited. In this case, current flow would be limited by the opening of the anti-crossing band gap associated with the condensate about the

Fermi level, at least up to some current and/or temperature at which the condensate would again collapse. This behavior, if not that required for BiSFET circuits, could at least provide an experimental signature of the condensate's existence.

Acknowledgments

This work was supported through the Nanoelectronics Research Initiative's (NRI's) Southwest Academy of Nanoelectronics (SWAN).

References

1. H. Min, R. Bistritzer, J.-J. Su, and A. H. MacDonald, *Phys. Rev. B* **78**, 121 401 (2008).
2. I. Sodemann, D. Pesin, and A.H. MacDonald, *arXiv*:1202nnnn (2012), submitted to *Phys. Rev. B*.
3. S. K. Banerjee, L. F. Register, E. Tutuc, D. Reddy and A. H. MacDonald, *IEEE Electron Device Lett.* **30**, 158-160, (2009).
4. ITRS.net.
5. I. B. Spielman, J. P. Eisenstein, L. N. Pfeiffer and K. W. West, *Phys. Rev. Lett.* 84, 5808-11 (2000).
6. J. P. Eisenstein and A. H. MacDonald, *Nature* **432**, 691-94 (2004).
7. D Basu, L. F. Register, A. H. MacDonald and S. K. Banerjee, *Physical Review B* **84**, 0354489 (2011).
8. D Reddy, L F. Register, E Tutuc, and S. K. Banerjee, *IEEE Trans. Electron Devices* **57**, 755-764 (2010).
9. D. Reddy, G. Carpenter, L F. Register and S. K. Banerjee, unpublished.
10. W. Jung and A. Hassibi, unpublished.
11. D. Basu, L. F. Register, D. Reddy, A. H. MacDonald and S. K. Banerjee, *Phys. Rev. B* **82**, 075409 (2010).
12. R. Bistritzer and A.H. MacDonald, *arXiv*:1002.2983v1 (2010).
13. X. Mou, L. F. Register, A. H. MacDonald and S. K. Banerjee, in preparation.

Current Switching in Crossed Graphene Nanoribbons

Roger K. Lake and K. M. Masum Habib

Department of Electrical Engineering, University of California Riverside, Riverside, CA 92521, USA

> Current switching by voltage control of the quantum phase has been demonstrated theoretically in crossed graphene nanoribbons. Notable features are the large suppression of the transmission over a wide range of energy of 1.2 eV at zero bias and the high sensitivity of the transmission to an applied bias, changing by 3 orders of magnitude with a bias change of 0.15 V. The magnitude of the current is modulated by a factor of 1000. The area of the intersection that is the active region of the device is 1.8nm x 1.8nm consisting of 120 atoms.

Introduction

Interest in twisted, or misoriented, layers of graphene was recently motivated by the need to understand the electronic properties of multilayer graphene furnace-grown on the C-face of SiC (1). Experimental analysis showed that the layers tended to be rotated with respect to each other at certain angles corresponding to allowed growth orientations with respect to the SiC substrate (2). Density functional theory calculations for such rotated bilayers found that the linear dispersion near the K-points was identical to that of single layer graphene (2). Thus, the rotation was found to be the cause of the decoupling, and the electronic states of the rotated layers of bilayer graphene were the same as those in isolated single-layer graphene. This picture was refined in a following study that found that for twist angles greater than ~3°, the low-energy carriers behave as massless Dirac Fermions and that for twist angles greater than 20°, the layers are effectively decoupled and act as independent layers of single-layer graphene (3). Theoretically, calculations of the density of states of randomly stacked graphene layers were first reported in 1992 (4). The more recent work has focused on calculating the energy-wavevector relation as a function of rotation angle using density functional theory (1,2,5,6,7), empirical tight binding (7,8), and continuum models (9). These studies agree that the misaligned graphene bilayers have a linear dispersion for any rotation greater than a few degrees. Very recently there have been studies of the effect of an applied vertical electric field (7,10). There has been one calculation of conduction between two rotated graphene sheets using a pi-bond model and a transfer Hamiltonian expression for current between the two layers (11). This study found the conductance to be enhanced at commensurate angles with relatively small unit cells.

Our work was initially motivated by a curiosity to understand the current-voltage (I-V) relation of two, independently contacted, overlapping, sheets of graphene (12,13,14). We initially considered two collinear armchair graphene nanoribbons (aGNR) 14 C-atoms wide (~1.8nm) with about the same length of overlap (12). The overlap region was arranged in both AB and AA stacking. The width was chosen as 3n+2 to minimize the bandgap. For these GNRs, the bandgap is 130 meV. As bias was applied, the I-V curve exhibited a peak and region of negative differential resistance.

We next considered two aGNRs crossed at right angles as shown in Fig. 1. In this configuration, the overlap region is neither AB nor AA stacking. For two infinite sheets, a 90° rotation is the same as a 30° rotation, which is not a commensurate rotation angle. The Moiré pattern can be observed at the intersection of the two nanoribbons in Fig. 1. The GNRs were terminated a few nanometers after the overlap region. In this crossed configuration, two current peaks were observed in the voltage region between 0 and 0.75 V. Furthermore, the peak-to-valley current ratios were rather large, 13 for the first peak and 7 for the second (14,15). This nonlinear I-V relation with multiple current peaks attracted the attention of circuit designers for applications in high-density functional circuits (15,16). The multiple current peaks were the result of the finite stub effects similar to stub effects in transmission line or waveguide theory. Electronic waves are reflected back from the cut ends giving rise to resonances and anti-resonances as a function of energy. Further, unpublished calculations by us show that the voltage spacing of the current peaks is decreased by increasing the length of the stubs. This is what one would expect viewing the finite ends as resonant cavities.

Infinite Crossed GNRs

However, we were ultimately interested in understanding the current-voltage relation of two infinite, crossed GNRs where finite length effects played no role. This configuration seemed closer to something that might be feasible to build experimentally. Numerically, we converted the finite ends into infinite leads by applying self-energies at all four GNR ends shown in Fig. 1 (numerical details are provided below). The physics governing the charge transport in this system is very different from the physics of the finite crossbars described above.

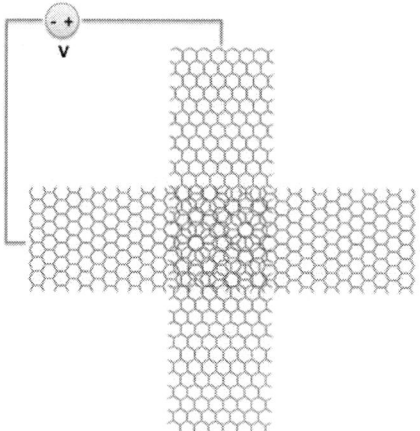

Figure 1. Graphene nanoribbon crossbar.

When the two infinite GNRs are at the same potential, they are effectively decoupled electronically. On a linear scale, the decoupling appears to be perfect. The eigenenergies at the bandedges at Γ ($k_x = k_y = 0$) are doubly degenerate. Plots of the two corresponding eigenstates show one with all of its weight on the top GNR and the other on the bottom GNR. The amount of decoupling can be better determined by looking at the transmission coefficient on a log scale as shown in Fig. 2. There are several things to note. The zero of

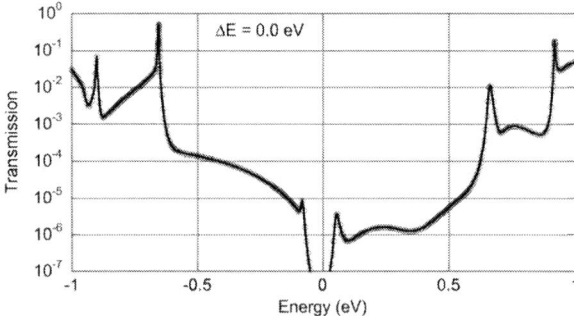

Figure 2. Transmission as a function of energy when the GNRs are at the same potential. All calculated points are shown. The Fermi level is set to E=0. The dip around E=0 corresponds to the 130 meV bandgap of the GNRs.

the energy scale corresponds to the Fermi level. The deep dip in transmission around E=0, corresponds to the 130 meV bandgap of the two GNRs. The transmission is strongly suppressed in the energy range of -0.7 eV < E < 0.7 eV. This is the energy range where there is only one mode propagating in each GNR. At E = ± 0.7 eV, the first excited mode turns on. Then there is significant coupling between the first-excited and the fundamental mode of the top and bottom GNRs. However, over a very large energy range of 1.6 eV (|E| < 0.8 eV), the transmission is suppressed by approximately 3 orders of magnitude. This suppression is the result of destructive quantum interference of the top and bottom wavefunctions. This was shown to be the decoupling mechanism for misoriented two-dimensional sheets of graphene (6), and we have explicitly demonstrated that destructive phase interference is the physical mechanism for the decoupling and suppression of the transmission for the crossed GNRs.

As the potential difference ΔE between the GNRs is increased, the transmission in the low-energy region increases rapidly with ΔE. Fig. 3 shows the transmission with ΔE = 150 meV. The transmission in the low energy region has increased by 3 to 4 orders of magnitude. Again, this is purely the result of the change in the phase of one wavefunction with respect to the other.

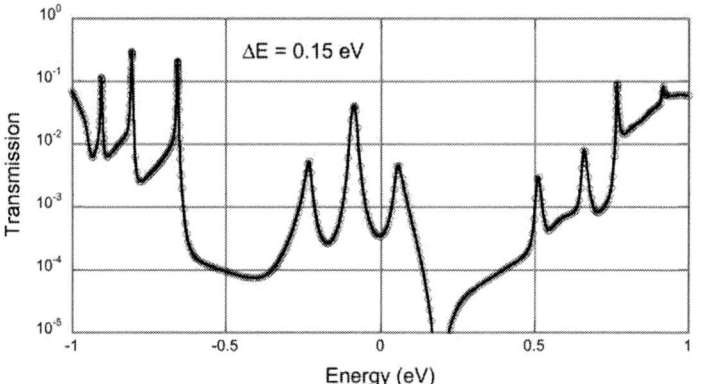

Figure 3. Transmisison when the potential of the top GNR is lowered by 0.15 eV with respect to the bottom GNR.

This sensitivity of the transmission to the energy difference of the two GNRs can be exploited with a built-in potential between the two GNRs. If we start with two GNRs that form a pn junction and apply a voltage as shown in Fig. 1, the applied bias will drive the two GNRs to equal potential. At that point the transmission will turn off. Calculation of the current voltage with an intial built-in potential of 0.25 eV results in the I-V curve shown in Fig. 4. At V=0.25 V, the potential between the two GNRs is driven to zero, and the current shuts off. The peak-to-valley current ratio is 1060.

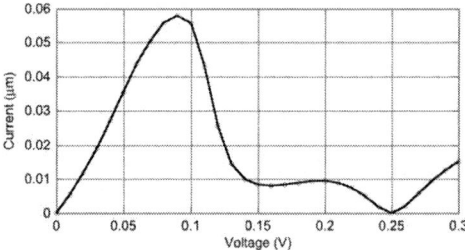

Figure 4. Current-voltage response of crossed graphene nanoribbons with a built-in potential of 0.25 eV.

As we mentioned above, we have explicitly shown that the mechanism of current switching is voltage control of the quantum phase of the wavefunctions. To show this we have calculated the current from a transfer Hamiltonian expression

$$I = \frac{4\pi e}{\hbar} \sum_{m,n} \int dE |M_{n,m}|^2 N_{1D}^m(E) N_{1D}^m(E-U) \left[f(E-\mu) - f(E-\mu-U) \right] \qquad [1]$$

where m and n designate the individual bands of the individual GNRs, N_{1D}^i is the 1D density of states of band i, f is the Fermi-dirac distribution, $U=eV$ is potential energy due to the bias, and $M_{n,m} = \langle \psi_n(k_y) | H_{int} | \psi_m(k_x) \rangle$ is the matrix element between the state

$|\psi_n(k_y)\rangle$ of the top GNR and the state $|\psi_m(k_x)\rangle$ of the bottom GNR. The matrix element can be decomposed into four components, $M = M_{AA} + M_{AB} + M_{BA} + M_{BB}$, where M_{AA} is the coupling between all the A atoms on the bottom GNR and all the A atoms on the top GNR and so on. The wavefunctions used for evaluating the matrix element are the analytical expressions derived in Ref. (17). We found that the interactions between the A atoms of one GNR with the B atoms of the other GNR are very weak compared to the interaction between the A(B) atoms of one GNR with the A(B) atoms of the other GNR. Thus, the value of the matrix element squared in Eq. [1] is solely a function of the M_{AA} and M_{BB} matrix elements. At zero bias M_{AA} and M_{BB} have equal magnitude but opposite sign making the resultant matrix element (M) very small. This leads to a strong suppression of the transmission at zero bias. This suppression is entirely the result of phase cancellation of the matrix element. At a higher bias, e. g., V= 0.25 V, the phase of M_{BB} is modified by the phase difference between the electronic wave functions of the B atoms of the individual GNRs. This phase difference is caused by the external bias. As a result, the two terms M_{AA} and M_{BB} no longer cancel, the total matrix element is no longer small, and the transmission is no longer suppressed. This analysis confirms that current switching is the result of voltage control of quantum phases.

Numerical Methods

Density functional theory as implemented in FIREBALL combined with a non-equilibrium Green function algorithm were used to calculate the transmission and I-V curves in Figs. 2-4. The details are the same as those described in Ref. (12). In calculating the I-V curve in Fig. 4, all of the applied potential is assumed to drop between the two GNRs. This makes sense from the point of view of a resistive voltage divider. The transmission of an ideal GNR is 1.0, whereas the transmission of the crossed GNRs in the region near the Fermi level is several orders of magnitude lower. Thus, the inter-GNR transfer is the rate-limiting step in the current flow, so that the crossing point is where the voltage will drop. We have repeated these calculations with a Huckel model that reproduces all of the features seen in the DFT results.

Conclusion

We have demonstrated current switching by voltage control of the quantum phase. Notable features are the large suppression of the transmission over an energy range of 1.2 eV at zero bias, and the high sensitivity of the transmission to an applied bias, changing by 3 orders of magnitude with a bias of 0.15 V. The current is modulated by a factor of 1000. The area of the intersection that is the active region of the device is 1.8nm x 1.8nm consisting of 120 atoms. This length scale is below any scale forseen in the ITRS Roadmap.

Acknowledgments

This work is supported by the Microelectronics Advanced Research Corporation Focus Center on Nano Materials (FENA).

References

1. X. Wu, X. Li, Z. Song, C. Berger, and W. A. de Heer, *Phys. Rev. Lett.*, **98**, 136801 (2007).
2. J. Hass, F. Varchon, J. E. Millán-Otoya, M. Sprinkle, N. Sharma, W. A. de Heer, C. Berger, P. N. First, L. Magaud, and E. H. Conrad, *Phys. Rev. Lett.*, **100**, 125504 (2008).
3. A. Luican, G. Li, A. Reina, J. Kong, R. R. Nair, K. S. Novoselov, A. K. Geim, and E. Y. Andrei, *Phys. Rev. Lett.*, **106**, 126802 (2011).
4. J.-C. Charlier, J.-P. Michenaud, and Ph. Lambin, *Phys. Rev. B*, **46**, 4540 (1992).
5. S. Latil, V. Meunier, and L. Henrard, *Phys. Rev. B*, **76**, 210402(R), (2007).
6. S. Shallcross, S. Sharma, and O. A. Pankratov, *Phys. Rev. Lett.*, **101**, 056803 (2008).
7. L. Xian, S. Barraza-Lopez, and M. Y. Chou, *Phys. Rev. B*, **84**, 075425 (2011).
8. S. Shallcross, S. Sharma, E. Kandelaki, and O. A. Pankratov, *Phys. Rev. B*, **81**, 165105 (2010).
9. J. M. B. Lopes dos Santos, N. M R. Peres, and A. H. Castro Neto, *Phys. Rev. Lett.*, **99**, 256802 (2007).
10. E. S. Morell, P. Vargas, L. Chico, and L. Brey, *Phys. Rev. B*, **84**, 195421 (2011).
11. R. Bistritzer and A. H. MacDonald, *Phys. Rev. B*, **81**, 245412 (2010).
12. K. M. M. Habib, F. Zahid, and R. K. Lake, *Appl. Phys. Lett.*, **98**, 192112 (2011).
13. K. M. M. Habib, S. Ahsan, and R. K. Lake, *Proc. SPIE,* **8101**, 81010Q (2011).
14. K. M. M. Habib and R. K. Lake, p. 109, *69th Annual Device Research Conference (DRC)* (2011).
15. K. M. M. Habib, A. Khitun, A. A. Balandin, and R. K. Lake, p. 86, *2011 IEEE/ACM International Symposium on Nanoscale Architectures (NANOARCH)* (2011).
16. S. Khasanvis, K. M. M. Habib, M. Rahman, P. Narayanan, R. K. Lake, and C. A. Moritz, p. 189, *2011 IEEE/ACM International Symposium on Nanoscale Architectures (NANOARCH)* (2011).
17. H. Zheng, Z. F. Wang, T. Luo, Q. W. Shi, and J. Chen, *Phys. Rev. B*, **75**, 165414 (2007).

CHAPTER 2

PERFORMANCE AND METROLOGY OF GRAPHENE

Device Characteristics of In-Situ CCVD Grown Bilayer Graphene FETs at Elevated Temperatures

Pia Juliane Wessely[1], Frank Wessely[1], Emrah Birinci[1], Bernadette Riedinger[2], Udo Schwalke[1]

[1]Institute for Semiconductor Technology and Nanoelectronics (ISTN), Technische Universität Darmstadt, Schlossgartenstrasse 8, 64289 Darmstadt, Germany
[2]Fraunhofer-Institut für Werkstoffmechanik, Wöhlerstrasse 11, 79108 Freiburg, Germany

> In this paper we report on transfer-free fabrication of bilayer graphene field effect transistors (BiLGFETs) on oxidized silicon wafers. By means of catalytic chemical vapor deposition (CCVD) the in-situ grown BiLGFETs are realized directly on oxidized silicon substrate, whereby the number of stacked graphene layers is determined by the selected CCVD process parameters, e.g. temperature and gas mixture. BiLGFETs exhibit ultra-high on/off-current ratios of 10^7 at room temperature, exceeding previously reported values by several orders of magnitude. The transfer characteristic shows a pure unipolar p-type device behavior. Furthermore, when increasing the ambient temperature to 200°C, the on/off-current ratio only degrades by one order of magnitude for BiLGFETs. Besides the excellent device characteristics, the complete CCVD fabrication process is silicon CMOS compatible. This will allow a simple and low-cost integration of graphene devices for nanoelectronic applications in a hybrid silicon CMOS environment.

Introduction

A monolayer of graphene consists of carbon atoms which are arranged in a quasi planar honeycomb lattice structure. This 2D material was exfoliated from graphite for the first time in 2004 by A. Geim and K. Novoselov [1]. However, size and position of the graphene flakes varies randomly and adsorbed molecules like O_2 and H_2O often accumulate at the interface between graphene and the substrate surface [2].

Currently there are several different manufacturing methods to produce graphene [3]. One possibility to produce graphene films is the use of CVD based methods to grow graphene on metallic substrates like copper or nickel [4]. Very recently a modified CVD-based approach has been reported which relies on the scalable synthesis of graphene on patterned Ni-dots [5]. Nevertheless, after the growth of the graphene sheets on the Ni dots it is still necessary to transfer the graphene layers. Although large area of graphene films can be produced [6] the disadvantage for integrated electronic applications is the need to transfer and align the produced graphene layer to the silicon substrate, for example [7].

Epitaxial graphene grown on silicon carbide (SiC) has been proposed by de Heer and Berger [8] in order to avoid graphene transfer. Using this method fairly large graphene sheets can be realized on a SiC wafer without the need to transfer. However, when comparing with conventional silicon processing, this method is more expensive because of the SiC substrate. Furthermore, the process requires extraordinary high growth

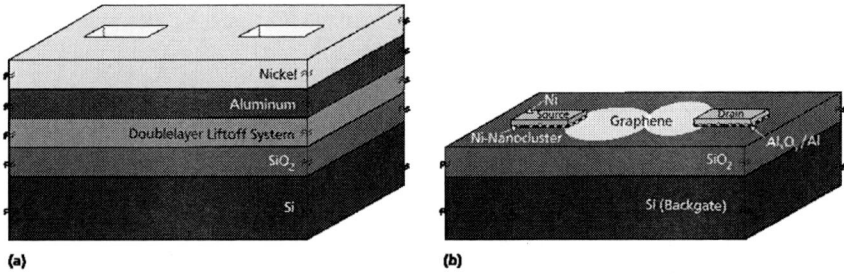

Figure 1: Schematic drawings illustrating the fabrication process: (a) Liftoff system used to pattern the catalyst areas. (b) Graphene FET structure produced by CCVD using an aluminum/nickel catalyst system. Note that the catalyst areas are simultaneously used as source/drain contacts.

temperatures of about 1400°C and is therefore not compatible with conventional silicon CMOS processing. Although graphene growth on dielectrics has been demonstrated by various research groups [9, 10, 11, 12, 13, 14], none of these research groups is growing graphene or graphite directly on thermally grown SiO_2 using conventional silicon substrates. Some of the materials are either extremely expensive or completely incompatible with CMOS process technology.

In order to avoid the above mentioned drawbacks we have developed a dedicated in-situ CVD-based growth method for graphene on oxidized silicon wafers. First experimental evidence demonstrating the feasibility of this transfer-free graphene growth method has already been published in November 2009 [15], [16]. Using a metallic catalyst seed, mono-, bi- and fewlayer graphene films are growing directly on SiO_2 covered Si wafers at moderate growth temperatures of 800 - 900°C by means of catalytic chemical vapor deposition (CCVD) from a methane feedstock [17]. Furthermore, by adjusting the CCVD process conditions we can adjust the number of grown graphene layers giving us the unique capability to investigate the physical and electrical properties of various in-situ grown graphene films at the device level [18, 19].

Fabrication

In preparation for CCVD a silicon wafer is oxidized in dry ambient at 1000°C for 120 min to obtain a 100nm thick SiO_2 film. Afterwards several lithography steps follow and a structured liftoff system remains on the wafer surface. Thin aluminum and nickel layers (5nm to 15 nm each) are evaporated over the whole substrate surface (c.f. Fig 1a) and are structured via liftoff. By annealing the wafer at 800°C to 900°C for 3 to 15 minutes in inert atmosphere, the aluminum transforms itself partially into aluminumoxide-like insulator (Al_xO_y) while the nickel (Ni) layer generates several nickel nanoclusters at the border of the catalyst system. In the subsequent methane-based CCVD process, graphene layers are growing on top of the silicon dioxide surface (c.f. Fig. 1b), while the number of the stacked graphene layers depends on the adjusted process parameters in particular process time and temperature. The methane flow rates are typically in the range of 4 to 15 litres per minute while the methane can be diluted by hydrogen with a flow rate of 3 litres per minute at maximum at atmospheric pressure. The

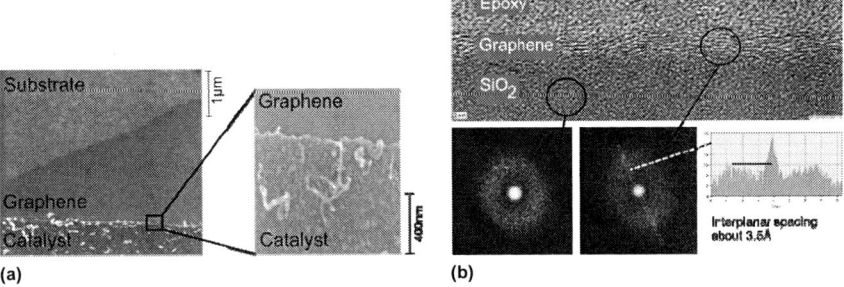

Figure 2: (a) HR scanning electron microscopy image of the probe at the catalyst graphene junction (b) Structural transmission electron microscopy examination of a multilayer graphene sample on a silicon dioxide surface with Fourier analysis.

total processing time for the wafers within the CVD chamber (Applied Materials AMV 1200) is in the range of 30 to 60 minutes [20]. During growth the graphene layer extends a few microns from the catalyst onto the oxidized wafer surface which is sufficient for device fabrication. When using a suitable device layout, these in-situ grown graphene films can be used as back-gated field effect device material contacted directly via the catalytic source/drain (S/D) areas (cf. Fig. 1b) for electrical characterization. These graphene devices possess a well defined channel length in the range of 1.6μm to 5μm while the channel width varies randomly from approximately 0.1μm to several microns, depending on local growth conditions [19].

When this CCVD process is completed several hundred of BiLGFETs are fabricated simultaneously on each 2'' wafer and the BiLGFETs are functional directly after the CCVD growth process.

Results and Discussion

Depending on the process parameters either monolayer, bilayer or fewlayer graphene films grow with this transfer-free approach on the silicon dioxide surface. The scanning electron microscopy (SEM) image of a fewlayer graphene sample at the catalyst graphene transition can be seen in figure 2a. The surface of the catalyst is rough and some carbon nanotubes growing from the nickel cluster are visible, as expected [15, 16]. Transmission electron microscopy (TEM) has been used in order to determine the inter-planar spacing between the graphene layers in a fewlayer graphene sample. The results of a Fourier-analysis of the TEM data of a fewlayer graphene sample reveals the crystalline properties of the transfer-free grown graphene multilayer more in detail (c.f. Fig 2b) [17]. In fact, the observed inter-planar spacing of 3.5Å is an additional strong evidence for the existence of fewlayer graphene grown by means of CCVD. Additionally, Raman spectroscopy of monolayer, bilayer and fewlayer graphene has been performed within the channel region (c.f. Fig 1b) of the graphene devices as already discussed in [19]. Figure 3a shows the Raman spectrum of in-situ CCVD grown bilayer graphene within the channel region of a BiLGFET. The Raman-measurements are performed using a Renishaw spectrometer at 633nm at room temperature. Comparing these results with [21]

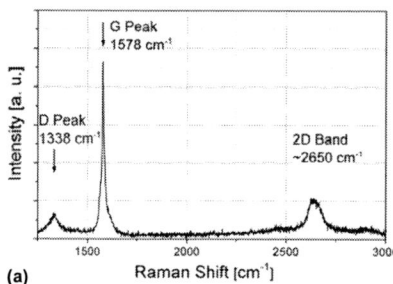

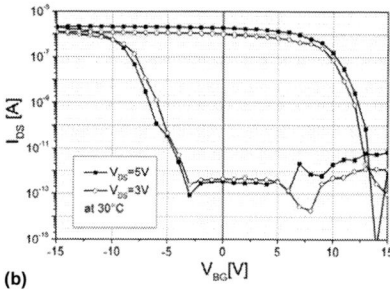

Figure 3: (a) Raman spectrum of in-situ CCVD grown bilayer graphene within the channel region of a BiLGFET. The Raman-measurements are performed using a Renishaw spectrometer at 633nm at room temperature. Comparing these results with [21] suggests the presence of bilayer graphene.
(b) Current voltage characteristic of a BiLGFET (L = 5.5μm, W < 50μm) as a function of the applied backgate voltage V_{BG} exhibiting an on/off-current ratio of $1 \cdot 10^6$ at room temperature. A hysteresis is observed in case of forward and backward gate voltage cycling.

confirms the presence of bilayer graphene grown by this transfer-free approach.

The electrical characterization of BiLGFETs is performed using a Keithley SCS 4200 semiconductor characterization system. Figure 3b shows the current voltage characteristic of a BiLGFET as a function of the applied backgate voltage. The catalyst pads are simultaneously used as source and drain contacts. We assume that a bandgap is partly induced by the electric field [19] of the applied backgate voltage (V_{BG}). However, additional effects, like intensive interactions between bilayer graphene and silicon dioxide may further enhance the bandgap. Such intensive interactions are expected develop during the growth of the bilayer graphene on the silicon dioxide at moderate temperatures under well defined ambient conditions within a CVD chamber [19]. In fact, in a similar fashion graphene-substrate interactions have been used to explain the substrate-induced bandgap opening in epitaxial graphene [22]. Since substantial amounts of atomic hydrogen are generated from the decomposition of CH_4 during CCVD processing, it is likely that H-atoms adsorb on the graphene surface or may be incorporated within the graphene bilayer. As a result, effects on the electronic properties such as increasing the bandgap are expected [23]. All BiLGFETs show a clear unipolar p-type device behavior which is consistent with the output characteristic [19]. The selection of the carrier type (i.e. holes in this case) may be facilitated by additional doping and/or Schottky-barrier effects. Furthermore, in view of the intensive graphene-substrate interactions, we suspect that atomic hydrogen, which is also known to passivate interface traps in SiO_2 [24], plays an important role in the carrier type selection. The current voltage characteristic is measured at room temperature and at elevated temperatures up to 200°C. Figure 4a shows the current voltage characteristic of a BiLGFET depending on the applied backgate voltage while a constant voltage of $V_{DS}=3V$ between drain and source is applied. In this case the current voltage characteristic is shown for 5 different

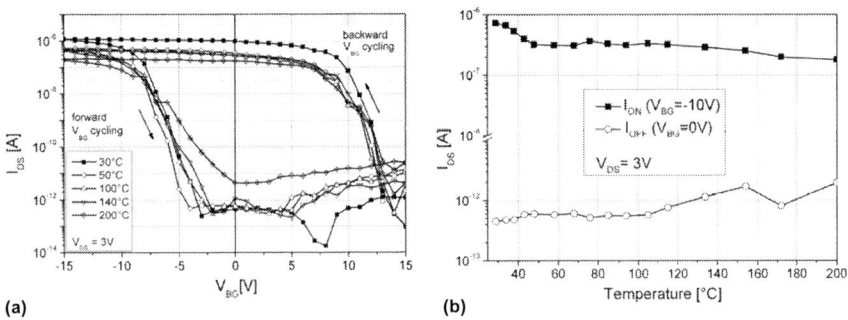

(a)   (b)

Figure 4: (a) Current voltage characteristic of a bilayer graphene transistor as a function of the applied backgate voltage V_{BG} at different temperatures. A hysteresis is observed in case of forward and backward gate voltage cycling. (b) BiLGFETs on-state current I_{ON}, defined for V_{BG} = -10V as well as the off-state current I_{OFF}, defined for V_{BG} = 0V, as a function of temperature.

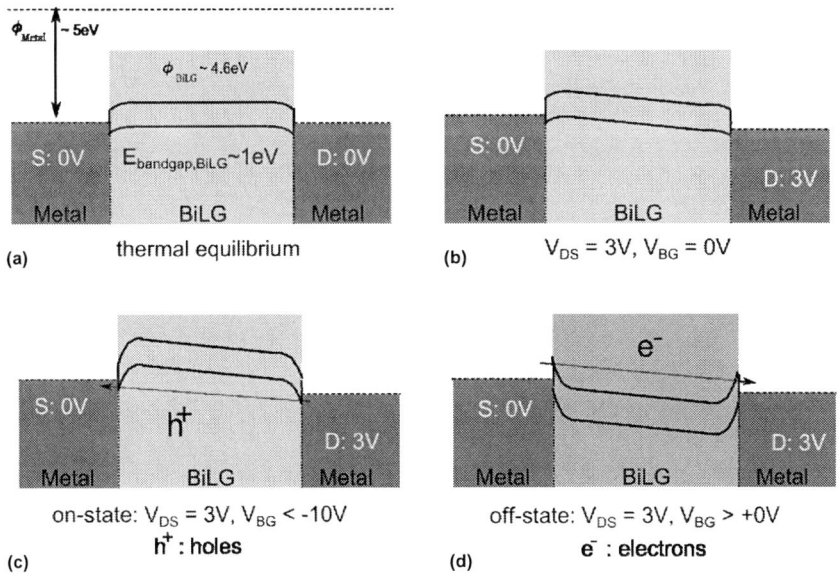

Figure 5: (a) Schematic band diagram for a BiLGFET in thermal equilibrium. (b) Schematic band diagram for a BiLGFET when a constant voltage of V_{DS}=3V is applied but V_{BG}=0V. (c) Schematic band diagram for a BiLGFET while conducting in the on-state, i.e. V_{BG} < -10V, only for the forward cycling range, i.e. V_{BG} is swept from -15V to 15V. (d) Schematic band diagram for a BiLGFET when in the off-state, i.e. V_{BG} > +0V, for the forward cycling range.

temperatures of 30°C, 50°C, 100°C, 140°C and 200°C (measured at the backgate contact). As seen in Figure 4a the hysteresis of this BiLGFET is mostly uniform for rising temperatures but decreases from ΔV_{BG} (30°C) = 16V over ΔV_{BG} (100°C) = 14V to ΔV_{BG} (200°C) = 13V. In-situ CCVD grown BiLGFETs exhibit an average hysteresis of $V_{BG,BiLGFET}$ = 19.5V ± 20% at room temperature [20]. Figure 4b shows the BiLGFETs on-state current I_{ON}, defined at V_{BG} = -10V as well as the off-state current I_{OFF}, defined at V_{BG} = 0V. I_{ON} decreases by factor two if the temperature is increased up to 200°C, whereby I_{OFF} increases less than one order of magnitude. The on/off-current ratio at 200°C is about $1 \cdot 10^5$ for the examined BiLGFET.

Schematic band diagrams for a BiLGFET are shown in Figure 5 in order to explain the experimental observations. A metal work function of approximately 5eV for nickel is assumed as well as a band gap of approximately 1eV for BiLG, deduced for comparison from carbon nanotubes showing a similar current voltage characteristic [15]. Previous studies [25, 26, 27, 28] reveal that the work function of graphene is in a similar range to that of graphite, ~4.6eV, [29] and depends sensitively on the number of layers [30, 31]. The schematic band diagram for a BiLGFET in thermal equilibrium is shown in figure 5a. Figure 5b shows the corresponding band diagram for a BiLGFET when a constant voltage of V_{DS}=3V is applied. The schematic band diagram for a BiLGFET in the on-state is shown in Figure 5c. In this drawing the conduction of holes is shown only for the forward cycling range, i.e. V_{BG} is swept from -15V to 15V. Therefore V_{BG} is smaller than -10V while a voltage between drain and source of V_{DS}=3V is applied. In this case electron conduction is suppressed due to the high barrier for electrons in consequence of band bending. Figure 5d shows the schematic band diagram for the off-state of a BiLGFET for the forward cycling range. Therefore V_{BG} is higher than +0V. Because of band bending the hole conduction is suppressed in this case. Electron conduction is possible in case of tunneling through the Schottky barrier as well as thermally assisted field emission.

The fact that the total decrease in the on/off-current ratio at 200°C is fairly small, approximately one order of magnitude, is partly attributed to the influence of the Schottky source/drain barriers. The increasing off-current is an evidence for an increasing probability of thermal emission of charge carriers over the Schottky barriers in case of increasing temperatures. However, the decrease of the on-current is related to a reduction in carrier mobility due to increased phonon scattering at elevated temperatures.

Conclusion

The examined BiLGFET exhibit an ultra-high on/off-current ratio of 1×10^6 at room temperature as well as an on/off-current ratio of 1×10^5 at 200°C exceeding previously reported values by several orders of magnitude. We explain the improved device characteristics by a combination of effects, in particular graphene-substrate interactions, hydrogen doping and Schottky-barrier effects at the source/drain contacts as well. With our transfer-free fabrication method hundreds of BiLGFETs have been realized simultaneously on one 2'' wafer by in-situ CCVD grown BiLG in a silicon CMOS compatible process.

Acknowledgments

This research is part of the ELOGRAPH project within the ESF EuroGRAPHENE EUROCORES program and partially funded by the German Research Foundation (DFG, SCHW1173/7-1).

References

1. K. Geim, K. S. Novoselov, *Nature Materials,* **6** (2007)
2. H. Wang, Y. Wu, C. Cong, J. Shang, T. Yu, *ACS Nano*, **4**, 12 (2010)
3. M. C. Lemme, *Solid State Phenomena* **156-158** (2010)
4. L. G. De Arco, Y. Zhang, A. Kumar, C. Zhou, *IEEE Trans. on Nanotechnology,* **8** (2009), 135-138
5. Y. Wang et al., *IEEE Trans. Electron Devices,* **57** (2010), 3472-3476
6. K. S. Kim et al., *Nature* **457** (2009), 706-710
7. A. Reina, X. Jia, J. Ho, D. Nezich, H. Son, V. Bulovic, M. S. Dresselhaus, J. Kong, *Nano Lett.,* **9** (2009), 30-35
8. W. A. de Heer et al., *Solid State Commun.,* **143** (2007), 92–100
9. C.Su, A.-Y. Lu C.-Y. Wu, Y.-T. Li, K.-K. Liu, W. Zhang, S.-Y. Lin, Z.-Y. Juang, Y.-L. Zhong, F-R Chen, L.-J.Li, *ACS Nano Letters,* **11** (2011), 3612-3616
10. A. Ismach, C. Druzgalski, S. Penwell, A. Schartzberg, M. Zheng, A. Javey, Y. Zhang, *Nano Letters,* **10** (2010), 1542-1548
11. L. G. de Arco, Y. Zhand, C. Zhou, *IEEE Trans. On Nanotechnol.,* **8** (2009), 135
12. Y. Miyasaka, A. Nakamura, J. Temmyo, *Japanese Journal of Applied Physics*, **50** (2011), 04DH12-1/4
13. M.H. Rümmeli, A. Bachmatuki, A. Scott, F. Börrnert, J. H. Warner, V. Hoffman, J.-H. Lin, G. Cuniberti, B. Büchner, *Nano Letters,* **4** (2012), 4206-4210
14. G. Lippert, J. Dabrowski, M. C. Lemme, C. Marcus, O. Seifarth, G. Lupina, *Phys. Status Solidi,* **11** (2011), 2619-2622
15. L. Rispal, U. Schwalke, *SCS Trans.,* (2009)
16. L. Rispal, P. J. Ginsel, U. Schwalke, *ECS Trans.,* **33** (9) (2010), 13-19.
17. P. J. Ginsel, F. Wessely, E. Birinci, U. Schwalke, *IEEE DTIS* (2011)
18. P. J. Wessely, F. Wessely, E. Birinci, U. Schwalke, *ECS Trans.,* (2011) in print
19. P. J. Wessely, F. Wessely, E. Birinci, K. Beckmann, B. Riedinger, U. Schwalke, *PhysicaE,* (2012), in print
20. P. J. Wessely, F. Wessely, E. Birinci, K. Beckmann, B. Riedinger, U. Schwalke, *Electrochem. Solid-State Lett.,* **15** (4) K1-K4 (2012), in print
21. A.C. Ferrari, J.C. Meyer, V. Scardaci, C. Casiraghi, M. Lazzeri, F. Mauri *Phys. Rev. Lett.,* **97** (2006), 187401
22. S. Y. Zhou, G.-H. Gweon, A. V. Fedrov, P. N. First, W. A. De Heer, D.-H. Lee, F. Guinea, A. H. Castro Neto, A. Lanzara, *Nature Materials,* **6** (2007)
23. R. Balog et al., *Nature Materials,* **9** (2010), 315-319
24. R. R. Radzouk, B. E. Deal, *J. Electrochem. Soc.: Solid-state Sci. Technoll,* **26** (1979), 1573-1581
25. Y.-J. Yu, Y. Zhao, S. Ryu, L.E. Brus, K.S. Kim, and P. Kim *Nano Lett.,* **9** (2009), 3430-3434
26. T. Filleter, K.V. Emtsev, Th. Seyller, R. Bennewitz, *Appl. Phys. Lett.,* **93** (2008), 133117.

27. S. S. Datta, D. R. Strachan, E. J. Mele, A. T. C. Johnson, *Nano Lett.,* **9** (2009), 7–11.
28. Y. Shi, X. Dong, P. Chen, J. Wang, L.-J. Li, *Phys. ReV. B,* **79** (2009), 115402.
29. T. Takahashi, H. Tokailin, T. Sagawa, *Phys. ReV. B* **32** (1985), 8317-8324.
30. G. Giovannetti, P.A. Khomyakov, G. Brocks, V.M. Karpan, J. Van den Brink, P. J. Kelly, *Phys. ReV. Lett.,* **101** (2008), 026803.
31. H. Hibino, H. Kageshima, M. Kotsugi, F. Maeda, F.-Z. Guo, Y. Watanabe, *Phys. ReV. B,* **79** (2009), 125437.

ECS Transactions, 45 (4) 31-37 (2012)
10.1149/1.3700450 ©The Electrochemical Society

Surface Potential Measurements of Reconfigurable p-n Junctions in Graphene

Yunfei Wang and Robert E. Geer

College of Nanoscale Science and Engineering, University at Albany, SUNY, Albany, New York 12203, USA

> The direct surface potential measurement of reconfigurable graphene p-n junctions in exfoliated graphene deposited on a buried split-gate test structure is reported. The experimental geometry permitted direct imaging of the Fermi level variation across the electrostatically-doped p-n junction using Kelvin Probe Force Microscopy (KPFM). The measured graphene p-n junction doping profile and junction width are in good agreement with predictions from finite-element calculations.

Introduction

Manipulation and control of electron current in a graphene p-n junction (e.g. electron waveguiding, reflection, focusing) is directly determined by the spatial gradient of the Fermi level across the junction (1-4). Sharp Fermi level gradients are associated with negative index 'lensing' of electrons in graphene while broader gradients are predicted to form reflective boundaries (4). Quantitative metrology of the Fermi level gradient at p-n junctions (induced either chemically or electrostatically) is absolutely essential to determine device performance, validate models for device design and switch architectures, and quantitatively determine the impact of defects on device function and leakage. Direct measurements of the Fermi level of back-gated graphene devices using Kelvin probe force microscopy (KPFM) have been previously published (5). However, due to the device geometry, it was not possible to directly investigate the surface potential in the junction region between electrostatically p-doped and n-doped regions due to the use of top-side gate electrodes. Using a novel split-gate test structure fabricated at the College of Nanoscale Science and Engineering (CNSE) a p-n junction was formed in exfoliate graphene entirely through buried gates. This structure enabled top-side KPFM to investigate the surface potential profile associated with electrostatic doping. Measured surface potential profiles from p-n junctions in exfoliate graphene were compared to simulations based on finite-element modeling. Although preliminary, the simulated and measured profiles exhibit good agreement.

Experimental

The fundamental KPFM surface potential measurement approach and the novel split-gate test structure are shown in Fig. 1. When not in electrical contact the difference between the Fermi levels of the KPFM conducting tip and the sample (exfoliate graphene, in this case) reflects the distinct work functions of the two materials. Electrical contact between the tip and the graphene results in a corresponding contact potential difference in response to Fermi level equalization. On application of an AC potential to the tip at or near the cantilever resonant frequency a finite tip displacement will result due to a steady state, field-induced force which is proportional to the vertical gradient of the tip-sample capacitance. In conventional scanning KPFM a DC offset bias – equal and opposite to the contact potential difference between the tip and sample (graphene) – will minimize (ideally eliminate) the steady-state field-induced force (6, 7). In this manner the nulling offset bias is used to map the surface potential of the sample. Presuming the intrinsic work functions of the graphene and the tip remain constant throughout the experiment (i.e. no chemical modification), this measurement can be used to spatially map the variation of the Fermi level of the graphene as electrostatic doping is induced. In practice it is necessary to measure the sample topography in the absence of applied fields so that a differential spatial reference is established to maintain a constant separation between the tip and the graphene (so-called 'lift' height).

Since the KPFM null condition is based on an electric-field induced force from the material of interest, it is not possible to map structures shielded by a conductor or structures for which the capacitance gradient associated with neighboring conductors dominates that of the tip-sample capacitance. The latter issue is resolved by minimizing the tip-sample separation so that the associated capacitance gradient is dominant. The former requirement demands that any electrodes defining a p-n junction in graphene reside below the graphene. This necessitated the fabrication of a buried, split-gate test structure so that both electrodes used to form a p-n junction were in the form of 'back gates'. A cross-sectional schematic of this test structure and the reference voltages for the KPFM measurement are shown in Fig. 1b. Test structures were fabricated using conventional front-end integrated circuit processing. Buried electrodes were fabricated using polysilicon. The polysilicon layer was approximately 100 nm thick. Each electrode was 2 μm in width. The separation between electrodes was 200±20 nm. Silicon oxide (TEOS) served as the gate dielectric (100 nm thick). Overall dielectric thickness was chosen to facilitiate optical inspection of graphene deposition (8). Top-side electrodes for KFPM ground formation were deposited via e-beam lithography and metal liftoff directly on the graphene far (~ 100's of μm) from the gate regions. Exfoliated graphene was transferred onto as-fabricated test structures using a conventional adhesive approach. Care was taken to maintain graphene in as pristine a condition as possible in the area of KPFM imaging.

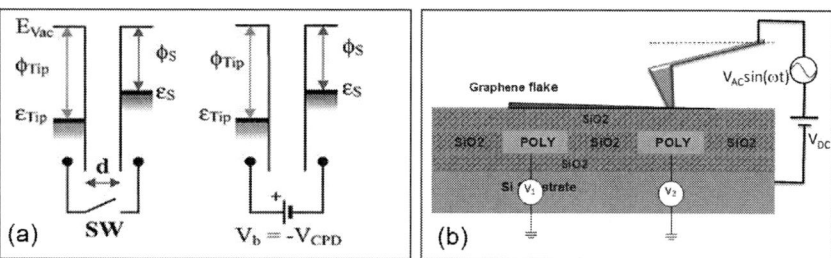

Figure 1. (a) (Left) Potential energy schematic for isolated KPFM tip and sample. (Right) Upon electrical connection of KPFM tip and sample a bias voltage is applied to null the contact potential difference. (b) Cross section schematic of buried, split gate test structure used for KPFM imaging of graphene p-n junctions. The AC and DC tip biases were referenced to ground.

Results and Discussion

KPFM Imaging of Graphene on Buried Split-Gate Test Structures. Figure 2a shows a 10μm × 10μm atomic force microscopy (AFM) scan of a 'stripe' of exfoliated graphene transferred to the split-gate test structure. The graphene exfoliate was approximately 2.5μm wide and 10μm in length. With the exception of the angled region at the lower terminus of the graphene 'stripe' and a thin (< 100nm) region on the leftmost edge, the exfoliate thickness was 1 monolayer as confirmed by Raman microscopy. The location of one pair of buried electrodes is also denoted in Fig. 2a. Figure 2b shows a 3-dimensional 10μm × 10μm KPFM map of the surface potential for the topographic region shown in Fig. 2a in the absence of gate bias (also with the location of the buried gates denoted). The contrast between the graphene and SiO_2 connotes the intrinsic difference between the graphene and SiO_2 work functions. Figure 2c shows a corresponding 3-dimension 10μm × 10μm KPFM image of the graphene exfoliate under application of 20V and -20V biases to the leftmost and rightmost buried gates, respectively. The corresponding modification of the surface potential associated with electrostatic doping (n-type for the left gate and p-type for the right gate) is evident in the image. The gate-induced surface potential shifts are restricted primarily to the graphene. As evident in Fig. 2c the measured graphene surface potential shows a substantial change between the two gate regions. The measured surface potential across a similar region of the SiO_2 dielectric adjacent to the graphene exhibits no such change. This confirms that the capacitance gradient between the tip and the graphene is the dominant component of the field-induced tip force used for the KPFM null point and that the KPFM profile shown in Fig. 2c originates from field-induced doping of the graphene rather than residual field effects from the buried gates. This is consistent with field-dependent Raman measurements on the same structure (not shown) (9).

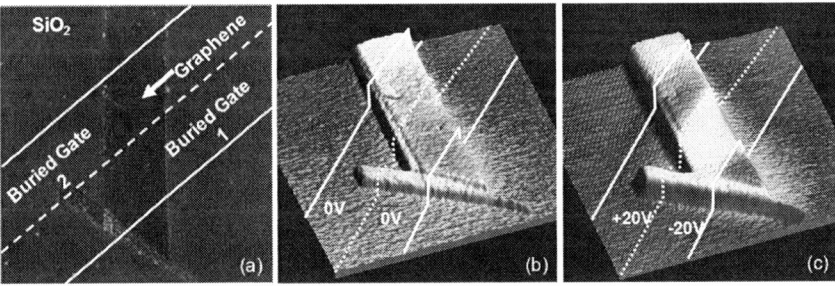

Figure 2. (a) 10μm × 10μm AFM topography scan of exfoliated graphene deposited on the split-gate test structure. The white lines denote the relative position of the two buried gates. (b) 10μm × 10μm KPFM surface potential image of the graphene exfoliate with both gates unbiased. (c) 10μm × 10μm KPFM surface potential image of the graphene exfoliate with split gates biased at ± 20 V.

Surface potential maps acquired upon switching the polarity of the gates are shown in Fig. 3. Figure 3a shows a 5μm × 5μm KPFM surface potential map of the graphene and SiO_2 in the absence of gate bias. Figures 3b and 3c show the corresponding 5μm × 5μm KPFM maps for energized gates with switched polarities. Note that the contrast scale for the KPFM image in Fig. 3c was shifted for clarity. In each panel of Fig. 3 the solid white line denotes the leftmost edge of the gate pair and the dashed white line denotes the gap between the split gates. The relative contrast of the graphene under switched gate polarity is clear. Note that the KPFM contrast transition in Figs. 2b and 2c across the solid white line (from unbiased graphene to biased graphene) appears sharper than the transition from positively to negatively biased gates (dashed white line) as expected.

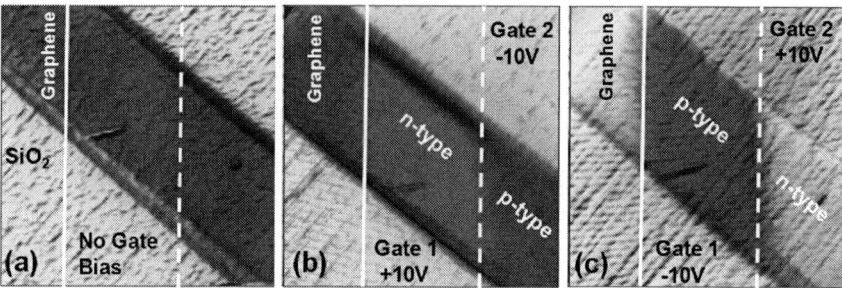

Figure 3. (a) Graphene surface potential (gates unbiased). (b) Graphene surface potential (V_{G1} = -10V; V_{G2} = +10V). (c) Graphene surface potential (V_{G1} = +10V; V_{G2} = -10V). The differential surface potential for panels (b) and (c) are actually the same. It appears lower due to grayscale conversion.

Finite Element Modeling of KPFM Surface Potential Profiles. Finite-element modeling of the surface potential measurements of exfoliated graphene was carried out for comparison with experimental measurements. Specifically, the simulations were undertaken to take into account the geometric effects associated with electrical coupling of the KPFM conducting tip within the local field geometry. The overall finite-element simulation configuration is illustrated in Fig. 4.

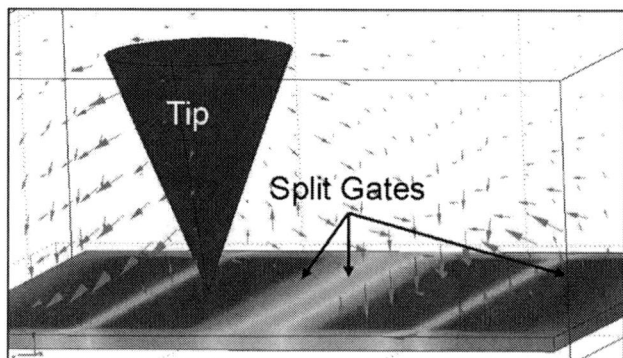

Figure 4. Schematic of simulation geometry. The KPFM probe is modeled as a conical tip (22.5° half-angle). Energized split gates are shown as alternating dark gray regions. The arrows denote the local electric field between the conducting tip and the (insulating) SiO_2 surface of the test structure.

The tip was modeled as a conducting cone with a 22.5° half-angle. A pyramidal tip model introduced unrealistic field gradients at the corners for reasonable mesh densities and was not used. The SiO_2 was modeled as an ideal dielectric ($\varepsilon/\varepsilon_0 = 4.0$) with a thickness of 100 nm.

Gate potentials were defined at the bottom side of the SiO_2 dielectric per the experimental conditions. The alternating potential at the topside of the SiO_2 dielectric for two pairs of energized gates is shown in Fig. 4. The arrows in that figure denote the local orientation of the electric field modeled in the absence of graphene. To simulate the KPFM measurement on graphene using the finite-element model shown in Fig. 4 the capacitance gradient associated with vertical displacement of the tip was calculated and used to determine the resultant electromagnetic force in the presence of an ideal graphene sheet (Fig. 5a). This force was minimized as a function of tip bias (Fig. 5b) for each position of the tip over the graphene at a specified lift height. For a tip-graphene separation of 10 nm the minimum of the field-induced electromagnetic force on the KPFM tip over an ideal monolayer graphene sheet subjected to a gate bias of 10V occurs at a tip bias voltage of approximately 0.25V. This approach simulates the null condition associated with the KPFM feedback mode and was used to model the surface potential profile across the graphene for comparisons with experimental measurements.

As expected, the surface potential simulated per this finite element model is sensitive to the lift height. A series of simulations was carried out as a function of lift height to estimate this effect. Based on extrapolation of these simulations it is estimated that a 10 nm KPFM tip lift height results in an apparent reduction of the surface potential by approximately 20%. The value of the experimental lift height of the KPFM tip is set by a tip-deflection offset. As such, it is subject to scan-to-scan variations and the overall calibration of the tip deflection. Calibration of the KPFM tip lift height for the measured data shown in Figs. 2 and 3 was in line with overall system calibration. The tip height during measurement was estimated at 10±3 nm. This translated to apparent surface potential reductions estimated at 10-30%. For experimental comparison the simulated surface potential was scaled by a constant factor within this range to match measured values at the edge of a KPFM scan.

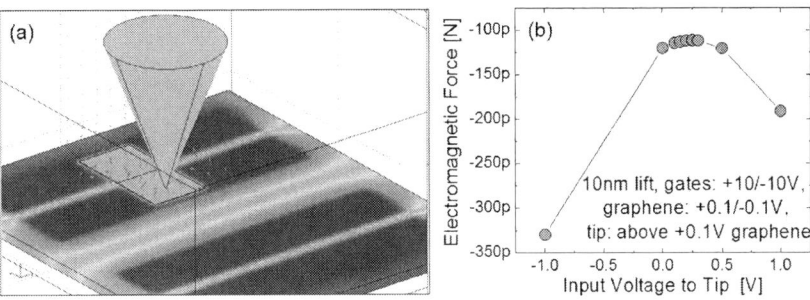

Figure 5. (a) Schematic of simulation geometry including the graphene exfoliate. The graphene field response was based on an ideal linear electronic dispersion relation. (b) Simulation of KPFM force null based on FE simulation assuming a 10 nm KPFM tip lift height. Simulated minimum force (null condition) occurs at approximately 0.25V tip bias.

Based on the finite element simluations described above, Fig. 6 shows a comparison between the measured graphene p-n junction profile and a finite element model (solid line). The circles denote KPFM measurements of the graphene exfoliate (Figs. 2 and 3) along the center of the 'stripe' across the junction region between the energized, buried gates. The 2nd solid line in Fig. 6 corresponds to a best fit of the experimental data to an error function (Gaussian-smeared step). The Gaussian width of this fit was 307±16 nm. The 2nd solid line in Fig. 6 corresponds to an error function representation of the simulated profile. This representation was used to take into account variations associated with the finite mesh size of the simulation. The Gaussian width of simulated junction profile was 243±8 nm.

Note that the overall potential offset between the measured and simulated profiles agrees well after scaling to correct for lift height. The measured Gaussian width of the p-n junction profile was approximately 64 nm larger than the finite element model prediction. Based on a preliminary analysis we speculate that this broadening results from field coupling between the sample and the tip+cantilever assembly that was not captured in the finite element simulations. Larger finite element models with finer mesh sizes are being developed to address this possibility.

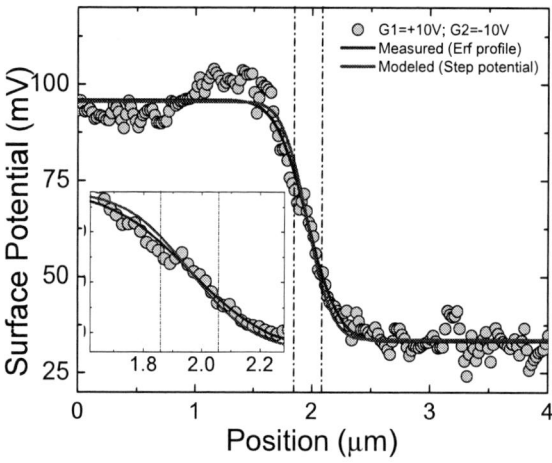

Figure 6. Measured (circles) and modeled surface potential profile across a graphene p-n junction. The 1st solid line is a best-fit of the measured profile to an error-function (Gaussian step) profile. The 2nd solid line corresponds to FE modeling results for the profile and agrees relatively well with the measured data. The best-fit Gaussian width is 64±12 nm broader than the FE modeled profile. The inset shows greater detail of the measured and simulated surface potential profiles across the junction region between the oppositely charged gates.

Conclusions

In conclusion, direct surface potential measurements of reconfigurable graphene p-n junctions have been carried out using via Kelvin Probe Force Microscopy (KPFM) of exfoliated graphene. Through utilization of a buried split-gate test structure it was possible to directly image the Fermi level variation across an electrostatically-doped p-n junction. Finite element modeling was carried out to compare with measured graphene surface potential profiles. The measured and simulated surface potential profiles of a graphene p-n junction exhibit good agreement. Additional modeling studies are underway to refine and improve the level of agreement.

Acknowledgments

Funding support for this work from the SRC/DARPA FCRP program and the SRC NRI program is gratefully acknowledged. We also are pleased to acknowledge Professor J. U. Lee and Mr. E. Comfort for fabrication of the split-gate test structure. Raman analysis for thickness determination of the graphene was provided by Dr. G. Rao.

References

1. V. V. Cheianov, V. I. Fal'ko, *Phys. Rev. B* **74**, 041403 (2006).
2. V.V. Cheianov, V. I. Fal'ko, B. L. Altshuler, *Science* **315**, 1252 (2007).
3. B. Huard, J. A. Sulpizio, N. Stander, K. Todd, B. Yang, and D. Goldhaber-Gordon, *Phys. Rev. Lett.*, **98**, 236803 (2007).
4. T. Low, S. Hong, J. Appenzeller, S. Datta, and M. S. Lundstrom, *IEEE Trans. Electron Devices*, **56** (6), 1292 (2009).
5. Y.-J. Yu, Y. Zhao, S. Ryu, L. E. Brus, K. S. Kim, and Philip Kim, *Nano Lett.* **9** (10), 3430 (2009).
6. M. Nonnenmacher, M. P. O'Boyle, and H. K. Wickramasinghe, *Appl. Phys. Lett.* **58**, 2921 (1991).
7. T Machleidt, E Sparrer, D Kapusi and K-H Franke, *Meas. Sci. Technol.* **20**, 084017 (2009).
8. Novoselov, K. S.; Morozov, S. V.; Mohinddin, T. M. G.; Ponomarenko, L. A.; Elias, D. C.; Yang, R.; Barbolina, II; Blake, P.; Booth, T. J.; Jiang, D.; Giesbers, J.; Hill, E. W.; Geim, A. K. *Physica Status Solidi B-Basic Solid State Physics* **244**(11), 4106 (2007).
9. Gayathri Rao, unpublished results.

38

In-situ Electrical Studies of Ozone based Atomic Layer Deposition on Graphene

S. Jandhyala[a], G. Mordi[b], B. Lee[a,*], J. Kim[a,†]

[a] Department of Materials Science & Engineering, The University of Texas at Dallas,
Richardson, Texas 75080, USA
[b] Department of Electrical Engineering, The University of Texas at Dallas, Richardson,
Texas 75080, USA
*Present address: Global Foundries, Malta, NY, 12020, USA.
†Corresponding author; E-mail: jiyoung.kim@utdallas.edu

> In this study, we present *in-situ* electrical studies of back-gated
> graphene field-effect transistors (GFETs) exposed to ozone at
> room temperature. Here, we compare the effect of the ozone
> exposure on graphene transport characteristics in two different
> environments using a vacuum probe station with static vacuum and
> an atomic layer deposition (ALD) chamber having a dynamic
> environment with continuous nitrogen purge. We observed that in
> both the cases there is a *p-type doping* in graphene upon exposure
> to ozone, but, there is a significant difference in the ozone
> exposure time required to witness a noticeable effect on the
> transport properties, with the dynamic environment requiring a
> longer ozone treatment period. The observed effect on the
> graphene devices was found to be a reversible phenomenon under
> vacuum suggesting a physisorption process for ozone adsorption
> on graphene. With the ozone functionalization approach, top-gate
> GFETs were prepared using chemical vapor deposited (CVD)
> graphene and ALD Al_2O_3 as gate dielectric with 10 nm thickness.

Introduction

One of the major advancements in the field of nano-electronics in the last decade was the
successful isolation of graphene [1] which led to the proposal of many novel electronic
and photonic devices [2-4]. Graphene is an allotrope of carbon that has attracted many
researchers from both academia and industry alike because of its interesting electronic
properties which arise from its Dirac cone like band structure [2-4]. Due to its two-
dimensional nature and exceptional properties such as high carrier mobilities, high
saturation velocity and long mean free path, it has been predicted to play a role not only
in future CMOS (complementary metal-oxide-semiconductor) applications, but also in
post-CMOS electronics. Although, graphene has a huge potential in future electronics
applications, there are a number of technological issues which need to be resolved. This
includes developing a technique for depositing high-quality dielectrics on graphene.
Deposition of dielectrics on graphene is important from the perspective of a gate insulator
for CMOS applications, as a tunneling barrier in case of spin-FETs and also for
passivation purposes. At the same time, the dielectric processes need to satisfy certain
criteria to be successfully implemented such as cause minimal change in graphene

properties, provide good interface characteristics electrically (low interface defects) and the thicknesses of the dielectrics are scalable for achieving a low-voltage operation.

Atomic layer deposition (ALD) is the technique of choice in the recent times for depositing thin and conformal high-κ dielectrics, but, graphene has a hydrophobic surface and an attempt to deposit oxides on it with the more conventional water (H_2O) based ALD process results in non-uniform films [4-6]. In order to overcome this problem, thin seed-layers (functionalization layers) such as evaporated metal or metal oxide layers, self-assembled monolayers (SAMs) or polymers [4] are introduced on graphene prior to the ALD process [4-5]. But, such techniques result in an unwanted low quality interfacial layer which limits the scalability of the dielectric thickness, which is required for greater control of the graphene channel in an FET configuration. Several other techniques for direct deposition of dielectrics such as physical vapor deposition (PVD) [5], plasma-enhanced chemical vapor deposition (PECVD) have also been explored [4,7]. But, such techniques either damage the graphene or result in unintentional doping in graphene.

In order to overcome these issues our group has developed a novel technique based on ozone (O_3) functionalization for depositing ALD oxides such as Al_2O_3 [6,8-10]. Top-gate graphene field-effect transistors (GFETs) fabricated using this technique displayed mobilities as high as 5000 cm^2/Vs [8,10]. In this technique, first, a seed-layer of Al_2O_3 is deposited using 6 cycles of TMA/O_3 at room-temperature (25 °C), followed by a higher temperature (200 °C) Al_2O_3 deposition using TMA/H_2O in order to achieve a conformal and pinhole free dielectric on graphene [8]. It has been predicted based on First-Principle calculations that O_3 can functionalize (adsorb on) the graphene surface either through a physisorption or a chemisorption process depending on the temperature and O_3 concentration [11]. In order to understand the mechanism of O_3 functionalization more carefully, the effects of O_3 on charge transport properties of graphene were investigated using back-gated GFETs by exposing them to O_3 and characterizing them *in-situ*. The observed charge scattering mechanisms and the effect on mobility due to the interaction of O_3 with graphene with respect to the type of environment are discussed. We also present electrical characteristics of top-gate GFETs prepared using CVD graphene and O_3 based ALD Al_2O_3 as gate dielectric with thickness of 10 nm.

Experimental Details

In case of the *in-situ* studies, back-gate GFETs were prepared using exfoliated graphene (obtained using natural graphite) on a ~290 nm thermal SiO_2/Si substrate. Optical microscopy and Raman spectroscopy (Nicolet Almega XR Raman system) were employed to identify single-layer graphene flakes. A 532 nm wavelength (green) laser with a power of ~1 mW was used for the Raman spectroscopy of the graphene flakes. Electron beam (e-beam) lithography was used to define the source/drain regions and the metal (5 nm Cr/50 nm Au) was deposited using an e-beam evaporation process. This was followed by a lift-off process in warm acetone (60 °C). The low-resistivity (0.005-0.01 Ohm-cm), p-type (Boron doped) Si substrate served as the back-gate electrode.

Two different setups were used for the *in-situ* studies and depending on the type of environment during ozone exposures; the system is defined as either a static environment (SE) or dynamic environment (DE). For SE studies, a cryo-probe station (Lake Shore) was modified in order to be able to introduce O_3 gas into the system. Figure 1 (a) shows the schematic of the setup. The back-gate GFETs prepared using the above described method were attached to the chuck (of the probe station) using a silver paste and then the chamber was pumped down to ~6 × 10^{-5} torr using a turbo pump. The system (chuck) was then heated to ~80 °C for 3 hours and then cooled to room-temperature (~25 °C) overnight before starting the experiments. This was done in an attempt to possibly remove any adsorbed atmospheric impurities or residues from the lithography process on the graphene surface. A solenoid valve was used for introducing small amounts of O_3 gas into the system. Before introducing O_3, the turbo pump was isolated using a manual valve and then an O_3 pulse was introduced into the system (under static vacuum). The chamber was then pumped down again using a roughing pump first and then with the turbo pump.

In case of DE experiments, a modified ALD system (PICOSUN SUNALE R-150) attached with an electrical feed through was used [see Figure 1(b)]. Back-gate GFETs prepared as described earlier were mounted on a gold coated package for wire-bonding the source, drain and back-gate electrodes in order to have electrical contact with the feed through extensions. In this case, the system was continuously purged with N_2 gas at a flow rate of 300 sccm controlled using a mass flow controller (MFC). The base pressure of the ALD system was ~0.75 torr.

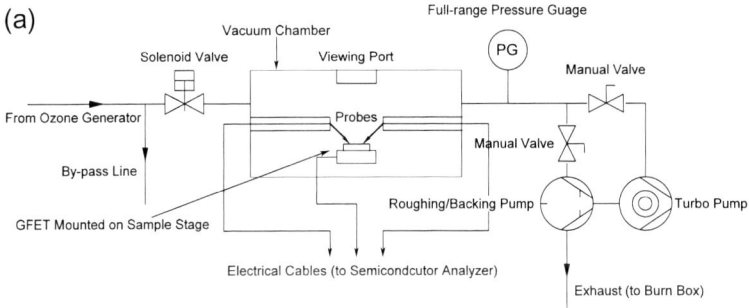

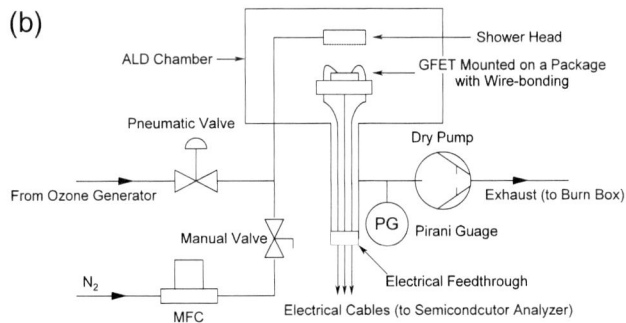

Figure 1. (a, b) Schematics of *in-situ* experimental setups for Static Environment (SE) experiments using a cryo-probe station and for Dynamic Environment (DE) using an ALD reactor.

In order to prepare top-gate GFETs, large area graphene obtained with chemical vapor deposition (CVD) technique on NiCu (alloy) films on Si substrate. CVD graphene was transferred was on to 300 nm SiO_2/Si substrate using a pressure sensitive adhesive ultraviolet tape. The details of CVD graphene growth and transfer are described elsewhere [12]. First, back-gate GFETs were prepared using a photolithography process, followed by an ALD process to deposit 10 nm Al_2O_3. The ALD process (Cambridge Nanotech Savannah 200) was carried using the room-temperature O_3 functionalization approach and trimethylaluminum (TMA) and water (H_2O) as precursor and oxidant respectively for depositing Al_2O_3. HF-last Si samples accompanying the graphene devices were used to confirm the thicknesses of the ALD oxide films (by ellipsometer measurements). The samples were then annealed in an Ar environment (1 atm) for 30 min to densify the oxide films. Finally, top-gate metallization was carried out using another step of photolithography. Here, a metal stack of 40 nm Ni/30 nm Au was used for source, drain and top-gate electrodes.

In all the experiments, remotely generated O_3 gas with a concentration of ~380 g/Nm^3 was used (TMEIC OP-250H-LT ozone gas generating system). All the electrical measurements were carried out at room-temperature (25 °C) using Agilent HP 4155A semiconductor analyzer and Agilent HP 4284 LCR meter. The top-gate GFETs were measured in air ambient using a Cascade probe station.

Results and Discussion

Figure 2 shows the effect of ozone exposure and subsequent pumping time on Dirac voltage (V_{Dirac}) or the minimum conductivity point and the mobility of back-gated GFETs for two different systems at room temperature (~25 °C). In case of the cryo-probe station or SE experiments, an O_3 pulse of 0.1 s was introduced into the chamber. This resulted in a shift in V_{Dirac} from 3 V towards a more positive voltage of 38 V and also the mobility of the devices dropped from ~7000 cm^2/Vs to 4200 cm^2/Vs [see Figure 2(a)]. The constant mobility model introduced by Kim et al. [13] was used for extracting the mobility of the GFETs by individually fitting the electron and hole branches. The effect of O_3 exposure seems to be reduced slowly under vacuum over an extended period of time eventually leading to complete recovery of the initial device properties. When back-gate GFETs were exposed to O_3 in an ALD system with a dynamic environment, we observe a similar effect except that longer O_3 exposure periods were required to see a change in the transport properties of the GFETs comparable to the SE experiments. Figure 2(b) shows that the shift in Dirac voltage is still small (ΔV_{Dirac} = 14 V) even after 30 minutes of O_3 exposure.

In separate studies based on *ab-initio* calculations, we have found that at 300 K (room temperature), O_3 tends to physisorb on graphene resulting in a positive shift in the Dirac point (E_D) [14]. This is due to the fact that O_3 physisorbed on the graphene surface acts as an acceptor and pushes the Fermi level (E_F) into the valence band resulting in *p-type doping* of graphene. It has been reported in graphene literature that dopants such as O_3 can act as charge scattering centers and can therefore the lower the carrier mobility in graphene [15], which explains the observed drop in mobility in GFETs upon O_3 exposure. But, when the samples exposed to O_3 are left in vacuum, O_3 starts to desorb from the graphene surface and the devices regain their initial characteristics.

Figure 3 shows the results for different O_3 exposure time dependent studies. It is observed that the drain current (I_{DS}) of the GFET (at V_{DS} = 10 mV, V_{BG} = 0 V) increases very slowly with time and does not saturate even after 10 minutes of O_3 exposure. The observed increase in drain current is because of the shift in V_{Dirac} as shown in the I_{DS}-V_{BG} measurements in Figure 3(b). We observe that there is no noticeable change in V_{Dirac} up until 10 s O_3 exposure (ΔV_{Dirac} = 2 V) almost two orders of magnitude difference compared to the SE experiments, where, even for 0.1 s O_3 exposure, we observed a Dirac voltage shift (ΔV_{Dirac}) of 35 V. Figure 3(c) shows the position of V_{Dirac} as a function of O_3 exposure time. Since the amount of physisorbed O_3 determines the shift in V_{Dirac}, the observed smaller V_{Dirac} shifts in case of the DE experiments suggests that diluting the O_3 with the carrier gas (N_2) and the continuous purging makes the adsorption process difficult. In the past there have been contradicting reports regarding the possibility of O_3 functionalization on graphene surfaces. It was reported by Lee et al. [6,8] that O_3 allows for depositing conformal ALD Al_2O_3 films on graphite/graphene, but it was observed by Pirkle et al. [16] that the surface condition plays a key role in Al_2O_3 deposition on graphite and it is difficult to deposit ALD oxide films on clean graphite surface (basal plane) using an O_3 process. We believe that even though graphite surface condition plays a role, the reason for the observed different results also arises out of the different environments (systems with different carrier/purge gas flow rates) used in these experiments, which seems to determine the amount of physisorbed O_3 as observed in our in-situ electrical studies.

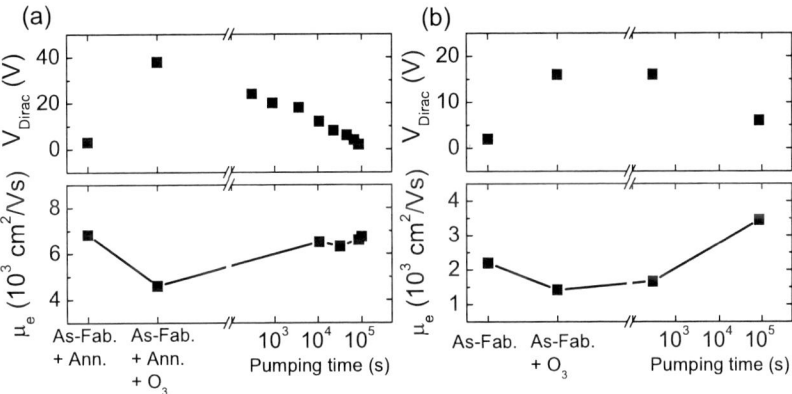

Figure 2. In-situ transport measurements of graphene devices with O_3 exposures. (a, b) Dirac voltage (V_{Dirac}) shifts and electron mobilities (μ_e) as a function of pumping time after O_3 exposure for SE and DE experiments respectively. The O_3 exposure time for SE and DE setups were 0.1 s and 30 min respectively.

One of the interesting things we had noticed in some of the samples was that even though a shift in V_{Dirac} was observed after O_3 exposure; there was no significant change in the mobility of GFETs, particularly in case of devices with a relatively lower mobility. But, when the samples were left in vacuum for a while, we observed an increase in mobility of GFETs [see Figure 4] along with the return of the V_{Dirac} close to the original value. We speculate that this is due to the cleanup of the graphene surface by O_3. It is often observed that graphene/graphite surface might have some atmospheric impurities adsorbed on its surface [16] and also the lithography process might leave some PMMA residues. Ozone might react with these impurities and can eventually remove them resulting in a cleaner surface and therefore enhanced device performance. In separate *in-situ* x-ray photoelectron spectroscopy (XPS) experiments it was observed that the atmospheric impurities adsorbed on graphite (HOPG) surface can be removed [17] or reduced upon exposure to O_3, which correlates with the observed electrical behavior of GFETs.

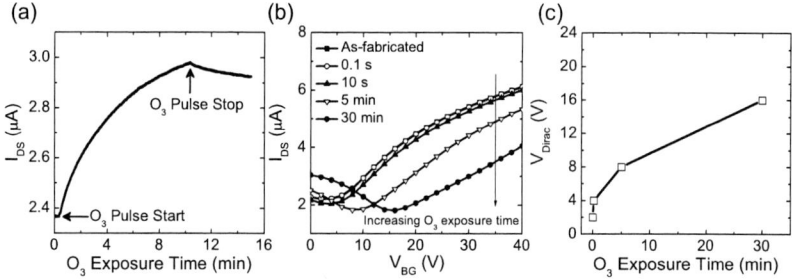

Figure 3. *In-situ* transport measurements of a graphene device with O_3 exposures in an ALD reactor. (a, b and c) Drain current (I_{DS}) at $V_{DS} = 10$ mV and $V_{BG} = 0$V, I_{DS}-V_{BG} curves and Dirac voltage (V_{Dirac}) shifts as a function of O_3 exposure time respectively. In Figure 3(b), the as-fabricated (line with solid squares) and 0.1 s O_3 exposure (line with open circles) curves overlap on each other.

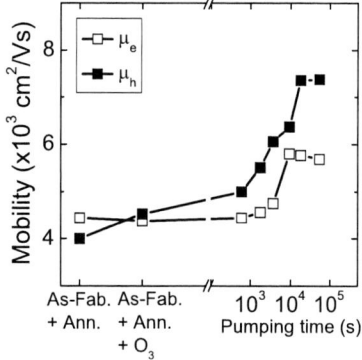

Figure 4. *In-situ* transport measurements of a graphene device with O_3 exposure (0.1 s) in the cryo-probe station. There is a clear increase in mobility when the sample is left under vacuum after ozone exposure.

It can be predicted from the above experiments that by controlling the partial pressure of O_3 ($p[O_3]$), the amount of O_3 adsorbed on graphene and therefore the nucleation site density for subsequent ALD deposition can be increased. An ALD deposition process utilizing a high-pressure O_3 functionalization approach was developed [14].

Figure 5 shows the two-terminal resistance (R_{Tot}) plots as a function of top-gate bias (V_{TG}) for a top-gate GFET with 10 nm ALD Al_2O_3 as gate-dielectric. It is observed that the operation voltage is reduced by decreasing the thickness of the oxide. The hysteresis width in these devices was ~0.2 V. Capacitance measurements of the top-gate GFETs were carried out using an ac bias of 100 mV at a frequency of 100 kHz overlaid on a dc bias applied to the top-gate electrode. A Graphene/10nm Al_2O_3/Metal (Ni/Au) capacitor shows non-linear C-V characteristics with 0.605 $\mu F/cm^2$ at V_{Dirac} (not shown). Since, in case of GFETs, quantum capacitance (C_Q) starts to dominate for thinner (or high capacitance) gate-dielectrics because of the small density of states in case of graphene, the gate capacitance (C_{TG}) does not directly correspond to the oxide capacitance (C_{ox}). So, we extracted the capacitance of the oxide layers using the equivalent capacitance circuit model [18]. The extracted oxide capacitance for 10 nm is 0.80 $\mu F/cm^2$, which corresponds to a dielectric constant of ~7.86. This is a reasonable value for low temperature Al_2O_3 films. It should be noted that the extracted oxide capacitances values are affected by the trapped charges at the interface which also results in the hysteresis observed in the resistance plots [see Figure 4] and capacitance measurements (not shown).

We also performed leakage current measurements on these devices (not shown) and found that the leakage current density levels were less than 10^{-4} A/cm^2 up to the break down field ~5-7 MV/cm.

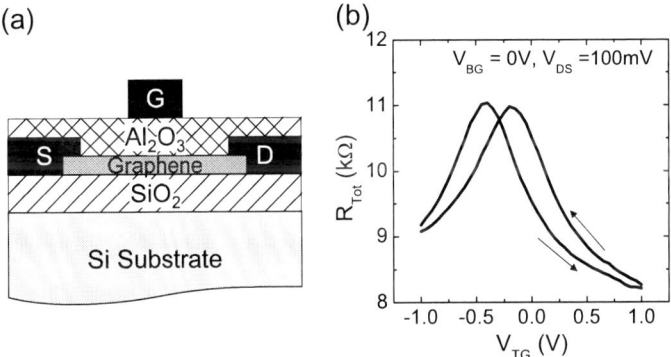

Figure 5. Electrical Characteristics of top-gate GFETs. (a) Schematic of top-gate GFETs used in this study with channel length (L_{Ch}) and width (W_{Ch}) being 5 μm and 10 μm respectively and the total length of the access region is 10 μm. (b) Two-terminal resistance (R_{Tot}) *versus* top-gate voltage (V_{TG}) plots for a GFET with 10 nm ALD Al_2O_3.

Conclusions

In summary we have used *in-situ* studies to investigate the effects of the ozone adsorption on graphene transport characteristics at room temperature and the effect of the environment type on the adsorption process. It is found that the O_3 exposure of graphene results in p-type doping and a drop in mobility of graphene, which can be reduced by leaving the samples under vacuum. Based on these observations it is claimed that the O_3 functionalization mechanism on graphene at room temperature is mainly a physisorption process and therefore the environment type used for O_3 functionalization process plays a key role. We have also shown that using the room temperature O_3 functionalization approach, thin and good quality gate dielectrics can be deposited on graphene, as evidenced by the extracted dielectric constants and leakage currents.

Acknowledgments

We would like to acknowledge the financial support provided by a grant(code #:2011K000211) from 'Center for Nanostructured Materials Technology' under '21st Century Frontier R&D Programs' of the Ministry of Education, Science and Technology (Korea), and SWAN and MIND programs funded through GRC-NRI. We would like to acknowledge TMEIC for providing the ozone generator (OP-250H-LT ozone gas generating system) and Drs. S. Seo and H. Chung of SAIT (Samsung Advanced Institute of Technology) for providing the CVD graphene used in this work. Useful discussions with Profs. K. Cho, R. M. Wallace, E. Vogel at UTD and Dr. L. Colombo at TI are acknowledged.

References

1. K. S. Novoselov, A. K. Geim, S. V. Morozov, D. Jiang, Y. Jiang, S. V. Dubonos, I. V. Grigirieva, and A. A. Firsov, *Science*, **306**, 666 (2004).
2. P. Avouris, *Nano Lett.*, **10** (11), 4285 (2010).
3. V. V. Cheianov, V. Fal'ko, B. L. Altshuler, *Science*, **315**, 1252 (2007).
4. S. K. Banerjee, F. L. Register, E. Tutuc, D. Basu, S. Kim, D. Reddy, A. H. MacDonald, *Proc. of the IEEE*, **98** (12), 2032 (2010).
5. L. Liao, X. Duan, *Mat. Sci. and Engg: R*, **70** (3-6), 354 (2010).
6. B. Lee, S.-Y. Park, H.-C. Kim, K. Cho, E. M. Vogel, M. J. Kim, R. M. Wallace, J. Kim, *Appl. Phys. Lett.*, **92** (20), 203102 (2008).
7. W. J. Zhu, D. Neumayer, V. Perebeinos, P. Avouris, *Nano Lett.*, 10, 3572 (2010).
8. B. Lee, G. Mordi, M. J. Kim, Y. J. Chabal, E. M. Vogel, R. M. Wallace, K. J. Cho, L. Colombo, J. Kim, *Appl. Phys. Lett.*, **97** (4), 043107 (2010).
9. G. Mordi, B. Lee, S. Jandhyala, J. Kim, *Proc. of 10th IEEE Int. Conf. on Nanotech. (NANO)*, 462 (2010).
10. S. Jandhyala, G. Mordi, B. Lee, J. Kim, *Proc. of IEEE Nanomat. and Dev. Conf. (NMDC)*, 94 (2011).
11. G. Lee, B. Lee, J. Kim, K. Cho, *J. Phys. Chem. C*, **113** (32), 14225 (2009).
12. J. Heo, H. J. Chung, Sung-Hoon Lee, H. Yang, D. H. Seo, J. K. Shin, U-In Chung, S. Seo, E. H. Hwang, S. Das Sarma, *Phys. Rev. B*, **84** (3), 035421 (2011).
13. S. Kim, J. Nah, I. Jo, D. Shahrjerdi, L. Colombo, Z. Yao, E. Tutuc, S. K. Banerjee, *Appl. Phys. Lett.*, **94** (6), 062107 (2009).
14. S. Jandhyala, G. Mordi, B. Lee, G. Lee, C. Floresca, P.-R. Cha, J. Ahn, R. M. Wallace, Y. Chabal, M. Kim, L. Colombo, K. Cho, J. Kim, Unpublished results.
15. J. H. Chen, C. Jang, S. Adam, M. S. Fuhrer, E. D. Williams, M. Ishigami, *Nature Phys.*, **4** (5), 377 (2008).
16. A. Pirkle, S. McDonnell, B. Lee, J. Kim, L. Colombo, R. M. Wallace, *Appl. Phys. Lett.*, **97** (8), 082901 (2010).
17. R. M. Wallace *et al.*, Unpublished results.
18. Z. Chen, J. Appenzeller, *Proc. of IEEE Int. Elec. Dev. Meeting (IEDM)*, (2008).

CHAPTER 3

GRAPHENE GROWTH AND CHARACTERIZATION

48

Direct Graphene Growth on Oxides: Interfacial Interactions and Band Gap Formation

J. A. Kelber[a], M. Zhou[a], S. Gaddam[a] and F. L. Pasquale[a]
L. Kong[b], and P. A. Dowben[b]

[a] Department of Chemistry, University of North Texas, Denton, Texas, 76203-5017 USA
[b] Department of Physics and Astronomy, University of Nebraska-Lincoln, Lincoln, Nebraska, 68588-0299, USA

> Direct graphene growth on oxides, without a metal catalyst, is a critical step towards the industrial-scale development of graphene devices. Such growth has been accomplished on MgO(111) by physical and chemical vapor deposition, and on $Co_3O_4(111)$ by molecular beam epitaxy. For MgO(111), there is evidence of significant oxide reconstruction and reaction with the first carbon layer, and a ~ 0.5-1 eV band gap, suggesting a commensurate graphene/oxide interface. For $Co_3O_4(111)$, growth occurs in a layer-by-layer fashion, with an incommensurate interface. In both cases, there is significant graphene-to-oxide charge transfer. The different results for the two different oxides appear to be due to the relative instability of the O-terminated (111) MgO surface. These results suggest a broad variety of potential electronic/spintronic applications, and indicate feasible routes towards fabrication of graphene devices by industrially practical and scalable methods.

I. Introduction

The direct growth of graphene on dielectric substrates by practical and scalable methods is a critical step towards the industrial scale development of graphene-based electronics and spintronics. Such methods include chemical and physical vapor deposition (CVD, PVD), and molecular beam epitaxy (MBE). To date, however, most graphene studies in the literature focus on either (a) physically-transferred graphene sheets from HOPG [1,2] or from graphene grown on transition metal substrates [3-5], or (b) the evaporation of Si from SiC(0001) [6-9]. The former method suffers from a number of practical difficulties, including the formation of nanoscale inhomogeneities [10,11]. The latter method appears limited to SiC. Over a decade ago, however, the growth of graphene/BN heterojunctions grown on transition metal substrates by chemical vapor deposition had begun to be explored [12-14]. More recently, the chemical vapor deposition (CVD) of graphene on h-BN(0001) grown by atomic layer deposition on Ru(0001) [15], and by CVD on h-BN(0001) nanoflakes [16] have lead to increased interest in deposition on nitrides [17].

In contrast, oxides have received relatively little attention as growth substrates until quite recently. Attempts at graphene growth by CVD on sapphire (0001) have resulted in the formation of graphene nanoflakes [18], although very recent preliminary accounts suggest that high-quality continuous sheets may be achieved at growth temperatures > 1800 K [19]. Graphene single and few layer films have been grown on MgO(111) by CVD of pre-dissociated C_2H_4 [20] and by magnetron sputter deposition at room temperature followed by annealing in UHV to induce ordering [21]. The few-layer graphene film grown on MgO(111) by CVD exhibited a band gap of ~ 0.5 eV - 1.0 eV, of considerable interest for logic device applications. Additionally, the direct controlled growth of single and few layer graphene on $Co_3O_4(111)$ by MBE has been demonstrated [22], with observation of C_{6V} symmetry in the low energy electron diffraction (LEED) pattern and an incommensurate oxide/graphene interface. MBE-induced graphene growth on mica [23] has also been reported, with Raman spectra suggesting a very low density of defects. A summary of results for direct graphene growth on oxide substrates is given in Table I.

Table I: Examples of direct growth of graphene on oxide substrates

substrate	growth temperature	method	remarks	reference
MgO(111)	~1000 K	CVD, PVD	Interfacial rehybridization, band gap	[20,21]
$Co_3O_4(111)$	1000 K	MBE	Incommensurate interface, no rehybridization	[22]
Mica	~1000 K	MBE		[23]
$Al_2O_3(0001)$	1800 K	CVD	High temperature required for few- defect films	[19]

As shown in Table I, growth temperatures are ~ 1000 K, except on alumina, and therefore compatible with Si CMOS processing. Notably, epitaxial growth at moderate temperatures is observed on substrates, with surface in-plane oxygen/oxygen nearest neighbor distances ~ 2.8 Å or more, presenting a lattice mismatch to the 2.5 Å graphene lattice constant. Thus, epitaxial growth of graphene on oxides is not critically dependent on a close lattice match.

The focus of this paper is on direct graphene growth on MgO(111) and on $Co_3O_4(111)$, as graphene growth on these oxides produces different results--probable interfacial reconstruction and observed carbon oxidation on MgO(111) [24], and an incommensurate interface and no evidence of carbon oxidation on $Co_3O_4(111)$. Additionally, in both cases there is evidence of significant graphene-to-oxide charge transfer [20,22,24], and this may be a general property of graphene growth on oxides.

II. Graphene growth on MgO(111)

Graphene has been deposited on MgO (111) single crystals by chemical vapor deposition, using C_2H_4 pre-dissociated by a UHV-compatible thermal cracker at 600 K, followed by annealing to 1000 K in UHV to induce long-range order [20]. Carbon deposition has also been carried out via magnetron sputter deposition (PVD) at room temperature from a graphite source, followed by annealing to 1000 K in UHV. The CVD process yielded a film of ~2.5 ML thickness, as determined by x-ray photoelectron spectroscopy (XPS) [20]. Formation of a monolayer of (111)-ordered carbon was also obtained by first reducing an overlayer of adventitious carbon to ~ 1 ML by mild heating in O_2, followed by annealing to 1000 K in UHV [21], as shown in Fig. 1. Core level binding energies for this system must be regarded with some skepticism, due to the extensive charging incurred for XPS studies on bulk samples of wide band-gap insulators. Regardless of the detailed binding energy, however, core level spectra for the ordered film show several C(1s) chemical environments, including a component ~ 3 eV higher binding energy than the main carbon feature (Fig. 1) and indicative of carbon in an oxidized state. As the data in Fig. 1 show, the evolution of this oxidized feature results directly from the thermally-induced reaction of the amorphous carbon overlayer with the oxide substrate (Fig. 1d), and coincides with the formation of long-range order, as evidenced by LEED (Fig. 1e,f).

Monolayer graphene on MgO(111)

The data in Fig. 1 also demonstrate that the evolved LEED pattern for a 1 ML film is actually of C_{3v}, rather than six-fold symmetry (Fig. 1f) [21]. This indicates that in the first layer, the chemical equivalence of the graphene A sites and B sites has been lifted, apparently due to reaction with the MgO substrate (Fig. 1d). This pattern of 3-fold symmetry is also observed for few-layer graphene on MgO(111) [21]. The fact that formation of an oxidized carbon component coincides with the formation of long range order and a C_{3v} LEED pattern argues strongly that the graphene/MgO interface is commensurate and involves both interfacial reconstruction and chemical reaction. Since the O-O nearest neighbor distance in bulk-terminated MgO(111) is ~ 2.8 Å [25], an incommensurate graphene/oxide interface will result if the oxide surface does not reconstruct. Carbon A sites and B sites would thus experience an ensemble of different substrate environments, resulting in equivalent the same average environment at both A and B sites. Instead, the 3-fold symmetry observed for single and few-layer films [21], coincident with the formation of an oxidized carbon peak (Fig. 1d) strongly suggests significant carbon and/or oxide reconstruction at the interface [24]. Indeed, this first layer may not be pure "graphene", but a partially oxidized, albeit ordered, form.

Charge transport data for a single layer C(111) film (produced by PVD) on MgO(111) are shown in Figs. 2 and 3. These data were acquired with conventional 6 probe geometry using evaporated Au contacts. A plot of ln(resistance) as a function of reciprocal temperature yields linear behavior (Fig. 2). A least-squares fit of the data indicates a carrier hopping activation energy of 0.64($\pm$ 0.05) eV. Current-voltage curves (Fig. 3) acquired at 200 K and at 300 K both show hysteresis, indicative of insulating behavior. Thus, spectroscopic and transport data strongly suggest that the first layer of "graphene" on MgO(111) is not true graphene, but an ordered, partially oxidized C(111) layer that has undergone a chemical reaction with the oxide substrate, and which displays semiconducting/insulating charge transport behavior.

Few-layer graphene/MgO:

Valence band photoemission and inverse photoemission spectra for a 2.5 ML film, produced by CVD of thermally dissociated C_2H_4, on MgO(111) are shown in Fig. 4 [20]. The corresponding LEED pattern is still of C_{3v} symmetry [21]. The angle-integrated valence band ultraviolet photoelectron spectrum (UPS, Fig. 4, left) closely resembles those of high quality graphene films on metallic substrates [26,27]. The k vector-resolved inverse photoemission spectrum (KRIPES, Fig. 4, right) shows the presence of π^* and σ^* features at relative energies corresponding to those reported for graphene films on SiC(0001) [9]. The data in Fig. 4 demonstrate an energy gap between the valence band maximum and the conduction band minimum. The magnitude of the band gap is ~ 0.5 eV - 1.0 eV, with uncertainty being due to both final state/substrate interactions in the UPS and KRIPES spectra, and the limiting energy resolution (0.4 eV) of the KRIPES measurements.

KRIPES, with monolayer sensitivity, also provides insight concerning the charge transfer between graphene and the substrate. Assuming simple band-filling, the energy difference between a given feature (e.g., the σ^* feature, Fig. 5) and the Fermi level provides a direct indication of the degree of band filling from or charge transfer to the substrate [20]. This is illustrated by the data in Fig. 5. The data for graphene/SiC(0001) (Fig. 5e) is due to multilayers of graphene (graphite) on SiC [9] and may be taken as an example of essentially zero charge transfer between graphene and substrate [20], as the "screening length" for eliminating charge transfer between graphene and the Si-terminated SiC substrate is ~ 1-2 graphene layers [28]. On this basis, the data in Fig. 5 demonstrate that graphene on BN/Ru(0001), and on Ru, Cu and Ni substrates, results in strong substrate-to-graphene charge transfer (n-type graphene doping). This is consistent with other results in the literature, particularly angle-resolved photoemission data for graphene/Ru(0001) [29]. In contrast, KRIPES data for graphene/MgO(111) indicate p-type doping, consistent with XPS data for this system [24]. The data therefore indicate that few layer graphene on MgO(111) exhibits photoemission and inverse photoemission spectra consistent with high quality graphene, but with a band gap of ~ 0.5-1eV. The data also demonstrate that graphene/oxide interactions result in strong graphene-to-oxide charge transfer.

III. Graphene on $Co_3O_4(111)/Co(111)$

Co(111) films were formed by MBE from a Co source onto a $Al_2O_3(0001)$ substrate at 750 K, followed by annealing in UHV to 1000 K [22]. This annealing resulted in segregation of dissolved oxygen to the surface and formation of a 3 ML film of Co_3O_4 (111), as determined by Auger and by LEED spectra in excellent agreement with the literature [30]. MBE onto the oxide from a graphite rod source at a sample temperature of 1000 K yielded the Auger data shown in Fig. 6. The C(KVV) Auger data (Fig. 6a-inset) indicate that carbon is sp^2 hybridized at all coverages observed, 0.4 ML-3 ML. The evolution of the C(KVV) intensity as a function of deposition time (Fig. 6 b) is well-described by a series of linear segments, with changes in slope corresponding to the completion of one monolayer and beginning of the next. This behavior is indicative of layer-by-layer growth [31]. LEED data (Fig. 7) indicate that MBE results in formation of an apparently six-fold pattern of diffraction spots; the outer ring of the bifurcated diffraction spots (Fig. 7a,c, and, e.g., G1 and G2 in Fig. 7b, d). These spots grow in intensity with increasing deposition time, whereas the other spots (e.g., arrows, Figs. 7a,c, and,e.g., O1, O2, Figs. 7b,d) decrease in intensity, clearly indicating that the outer ring of spots (line scans, Fig. 7b,d) are related to the growing carbon film. The carbon-related spots have a direct lattice spacing of 2.5 Å, relative to the $Co_3O_4(111)$ lattice spacing of 2.8 Å, which corresponds to the O-O nearest neighbor distance in the surface oxide layer [30]. Thus, the graphene/Co_3O_4 (111) interface is incommensurate. The Auger (Fig. 6) and LEED data (Fig. 7) demonstrate that MBE of carbon onto Co_3O_4 (111) at 1000 K yields sp^2 hybridized, (111)-ordered carbon--graphene. Raman data [22] and inertness of the sample upon exposure to ambient indicate that the 3ML graphene film is continuous and uniform over macroscopic distances. Spectroscopic ellipsometry data [22] also indicate that graphene/Co$_3$O$_4$(111) displays optical absorption and other electronic properties similar to graphene physically transferred to SiO$_2$, or grown thermally on SiC(0001). Thus, MBE from a graphite source onto Co$_3$O$_4$(111) results in the controlled layer-by-layer growth of macroscopically continuous and uniform graphene films, without significant reaction with or reconstruction of the oxide substrate.

A detailed LEED I(V) analysis [22] demonstrates that the graphene LEED pattern displays C$_6$v symmetry, in contrast to graphene/MgO(111) [21], but consistent with formation of an incommensurate graphene/oxide interface [24]. Further, the XPS C(1s) photoemission spectrum [22] displays an asymmetric characteristic of graphite/graphene and a π-to-π* transition satellite feature, but with a binding energy of 284.9(± 0.1) eV. Such a binding energy is 0.4 eV higher than that commonly observed for graphite, but close to the value reported--284.75 (±0.1 eV) [28]-- for single-layer graphene thermally grown on Si-terminated SiC(0001). Such a value is indicative of charge transfer from the carbon atoms to the substrate [32] in both systems, and as observed for graphene/MgO(111) [24]. The XPS data thus indicate strong graphene-to-oxide charge transfer on both MgO(111) and on Co$_3$O$_4$(111). In contrast to MgO(111), however, the XPS and other data provide no evidence of oxidized carbon or other interfacial chemical reaction at the graphene/Co$_3$O$_4$(111) interface.

The growth temperature resulting in ordered graphene growth, 1000 K, is similar to that reported for graphene on silica-terminated mica [23](Table I), but significantly lower than the minimum temperature required to obtain long range order during MBE on SiC, where only amorphous, sp^2 carbon is observed at deposition temperatures below 1273 K [33,34]. This indicates that carbon atom or nanocluster mobility on Co$_3$O$_4$(111) and on silica-terminated mica is greater at 1000 K than on SiC.

IV. MgO(111) vs. Co$_3$O$_4$(111): Why the difference?

Graphene deposition on both MgO(111) and Co$_3$O$_4$(111) results, in each case, in charge transfer from graphene to the substrate. However, in the case of growth on MgO(111), a band gap is observed for both single and few-layer samples (Figs. 3-5). Additionally, XPS yields evidence of interfacial reaction (Fig. 1b,d), resulting in some of the carbon being in a highly oxidized state. The formation of a band gap, together with a C$_3$v LEED pattern (Fig. 1f) strongly suggest a commensurate interface, with first row carbon atoms in distinctly different chemical environments for graphene "A" sites and "B" sites (Fig. 8a-c). This interaction lifts the degeneracy of the HOMO and LUMO orbitals at the Dirac point [35,36], resulting in a band gap (Fig. 8c). In contrast, graphene on an unreconstructed surface would result in both A and B graphene sites experiencing an ensemble of chemical environments (Fig. 8f), resulting in C$_6$v symmetry for the LEED pattern, and a retention of HOMO/LUMO degeneracy at the Dirac point and the absence of a band gap. The spectroscopy and transport data for graphene/MgO(111) (Figs 1-4) indicate band gap formation for both single and few-layer films, strongly suggesting the formation of a reconstructed, commensurate interface. Such a reconstruction is not unlikely, as the (111) planes of oxides with the rocksalt structure are entirely composed of either O anions or metal cations, resulting in an unstable surface energy [25]. Such structures are frequently stabilized in vacuum by surface termination with hydroxyl groups [37,38]. For this reason, MgO(111) and other (111) oxide surfaces with the rocksalt structure are prone to reconstruction upon metallization [39].

In contrast, graphene formation on $Co_3O_4(111)$, which has a spinel structure [30], yields an incommensurate interface (Fig. 7), with no evidence of oxide reconstruction, or of formation of a separate, higher oxidation state in the C(1s) XPS spectrum [22]. Based on the logic of Figs. 8 (d-f), no band gap formation is predicted for this system. The incommensurate interface and lack of reaction or reconstruction at the oxide surface is consistent with observations of the $Co_3O_4(111)$ surface structure, which is somewhat relaxed from the bulk-terminated structure to further reduce surface instability and the tendency to reconstruct [30]. Thus, the $Co_3O_4(111)$ surface is sufficiently stable to support graphene growth without reconstruction or chemical reaction with the adjacent carbon atoms.

V. Summary and Conclusions

The data presented here, and in recent publications [20-22] indicate that graphene layers can be deposited directly on MgO(111) (by CVD and PVD), and on $Co_3O_4(111)$ by MBE. Graphene formation on MgO(111) yields a band gap of ~ 0.5 - 1 eV, for both single and few layer films (Figs. 3-5). Indeed, the core level photoemission (Fig. 1) and transport data (Figs. 2,3) indicate that the first layer may not be true graphene, but a significantly oxidized form, albeit one with long-range order. Graphene has been grown directly on $Co_3O_4(111)$ in a controlled, layer-by-layer manner, up to a thickness of at least three monolayers [22] (Figs. 6 and 7). The LEED data (Fig. 7) show six-fold symmetry for both sub-monolayer and three-monolayer films, indicating an incommensurate oxide/graphene interface, which leads to the prediction of zero band gap for graphene/Co_3O_4(111) (Fig. 8).

The different properties of graphene on the two different oxides result in different potential applications. The size of the band gap for graphene/MgO(111) is suitable for field effect transistors (FETs) [40,41]. Since MgO(111) growth on Si(100) has been reported [42], formation of graphene-based FETs on Si(100), appears feasible. The magnetic polarizability of $Co_3O_4(111)$ is predicted to induce magnetic behavior in graphene conduction electrons--the proximity effect [43]. This, and the fact that $Co_3O_4(111)$ can be grown directly on Si(100) suggests a variety of spintronics applications for graphene on $Co_3O_4(111)$, and perhaps on the (111) surfaces of other magnetically polarizable oxides.

It should be pointed out that direct growth methods have only begun to be explored. PVD of graphene on MgO (111) was carried out at room temperature, followed by annealing to 1000 K in UHV. The direct growth of graphene on oxides by MBE would seem to be most promising, as this method has been demonstrated for mica [23] as well as for Co_3O_4(111). Indeed, a preliminary account [44] suggests that this approach may be successful for MgO (111) as well. There are obvious problems in depositing multilayer graphene films, even on metal surfaces, as once the first layer of graphene is complete, the growth surface becomes relatively inert [3]. Exceptions occur (e.g., Ni, Ru) for systems in which C has a relatively high solubility in the metal substrate [4,29], but this process is difficult to tailor to an arbitrarily-designated graphene thickness, and in any case appears limited to certain metal substrates. In contrast, the ability to grow few layer films in a controlled fashion by MBE (e.g., Fig. 6) argues that MBE may be quite practical for few layer films on dielectric substrates. Few layer films may afford a route toward the mitigation or "tuning" of effects due to substrate interactions, such as charge transfer or induced 3-fold graphene lattice symmetry and band gap formation [20,21]. This suggests that the relative orientations of adjacent graphene layers, and the resulting effects on electronic and magnetic properties, will become topics of considerable practical interest for device applications. The ability to grow graphene in a controlled manner on oxides at moderately elevated temperatures indicates that the practical development of graphene-based electronics and spintronics is now entirely feasible.

Acknowledgements

This work was supported by the Global Research Consortium of the Semiconductor Research Corporation under Task ID 2123.001. Jincheng Du is also gratefully acknowledged for providing a model of the incommensurate graphene/MgO interface, and for many enlightening discussions.

References

1. K.S. Novoselov, A.K. Geim, S.V. Morozov, D. Jiang, Y. Zhang, S.V. Dubonos, I.V. Grigorieva, A.A. Firsov, *Science,* **306,** 666 (2004).
2. K.S. Novoselov, A.K. Geim, S.V. Morozov, D. Jiang, M.I. Katsnelson, I.V. Grigorieva, S.V. Dubonos, A.A. Firsov, *Nature,* **438,** 197 (2005).
3. X. Li, W. Cai, J. An, S. Kim, J. Nah, D. Yang, R. Piner, A. Velamakanni, I. Jung, E. Tutuc, S.K. Banerjee, L. Colobmo, R.S. Ruoff, *Science,* **324,** 1312 (2009).
4. A. Reina, X. Jia, J. Ho, D. Nezich, H. Son, V. Bulovic, M.S. Dresselhaus, J. Kong, *Nanolett.* **9,** 30 (2009).
5. P.W. Sutter, J. Flege, E.A. Sutter, *Nature Materials,* **7,** 406 (2008).
6. C. Berger, Z. Song, T. Li, Z. Li, A.Y. Ogbazghi, R. Feng, Z. Dai, A.N. Marchenkov, E.H. Conrad, P.N. First, W.A. de Heer, *J. Phys. Chem. B,* **108,** 19912 (2004).
7. C. Berger, Z. Spong, X. Li, X. Wu, N. Brown, C. Naud, D. Mayou, T. Li, J. Hass, A. Marchenkov N., E.H. Conrad, P.N. First, W.A. de Heer, *Science,* **312,** 1191 (2006).
8. W.A. de Heer, C. Berger, X. Wu, P.N. First, E.H. Conrad, X. Li, T. Li, M. Sprinkle, J. Hass, M.L. Sadowski, M. Potemski, G. Martinez, *Sol. St. Comm.,* **143,** 92 (2007).
9. I. Forbeaux, J.- M. Themlin, J.- M. Debever, *Phys. Rev. B,* **58,** 16396 (1998).
10. Y. Zhang, V.W. Brar, C. Girit, A. Zettl, M.F. Crommie, *Nature Physics,* **5,** 722 (2009).
11. A. L. V. de Parga, F. Calleja, B. Borca, M. C. G. Passeggi Jr., J. J. Hinarejos, F. Guineau, R. Miranda, *Phys. Rev. Lett.,* **100,** 056807 (2008).
12. C. Oshima, A. Itoh, E. Rokuta, T. Tanaka, K. Yamashita, T. Sakuri, *Sol. St. Commun.,* **116,** 37 (2000).
13. C. Oshima, A. Nagashima, *J. Phys. : Cond. Matt.,* **9,** 1 (1997).
14. A. Nagashima, N. Tejima, Y. Gamou, T. Kawai, C. Oshima, *Phys. Rev. Lett.,* **75,** 3918 (1995).
15. C. Bjelkevig, Z. Mi, J. Xiao, P.A. Dowben, L. Wang, W. Mei, J.A. Kelber, *J. Phys.: Cond. Matt.,* **22,** 302002 (2010).
16. X. Ding, G. Ding, X. Xie, F. Huang, M. Jiang, *Carbon,* **49,** 2522 (2011).
17. M. Yang, C. Zhang, S. Wang, Y. Feng, Ariando, *AIP Adv.,* **1,** 03211 (2011).

18.] S.K. Jerng, D.S. Yu, Y.S. Kim, J. Ryou, S. Hong, C. Kim, S. Yoon, D.K. Efetov, P. Kim, S.H. Chun, *J. Phys. Chem. C,* **115,** 4491 (2011).
19. M.A. Fanton, J.A. Robinson, B.E. Weiland, M. LaBella, K. Trumbell, R. Kasarda, et al., *Abstract for the Graphene 2011 Conference* (Bilbao, Spain) (2011) .
20. L. Kong, C. Bjelkevig, S. Gaddam, M. Zhou, Y.H. Lee, G.H. Han, H.K. Jeong, N. Wu, Z. Zhang, J. Xiao, P.A. Dowben, J.A. Kelber, *J. Phys. Chem. C,* **114,** 21618 (2010).
21. S. Gaddam, C. Bjelkevig, S. Ge, K. Fukutani, P.A. Dowben, J.A. Kelber, *J. Phys.: Cond. Matt.,* **23,** 072204 (2011).
22. M. Zhou, F.L. Pasquale, P.A. Dowben, A. Boosalis, M. Schubert, V. Darakchieva, R. Yakimova, L. Kong, J.A. Kelber, *J. Phys.: Cond. Matt.,* **24,** 072201 (2012).
23. G. Lippert, J. Dabrowski, M. Lemme, C. Marcus, O. Seifarth, G. Lupina, *Phys. Stat. Sol. B,* **248,** 2619 (2011).
24. J.A. Kelber, S. Gaddam, C. Vamala, S. Eswaran, P.A. Dowben, *Proc. SPIE,* **8100,** 8100Y-1 (2011).
25. V.K. Lazarov, R. Plass, H.-. Poon, D.K. Saldin, M. Weinert, S.A. Chambers, M. Gajdardziska-Josifovska, *Phys. Rev. B,* **71,** 115434 (2005).
26. Y.S. Dedkov, M. Fonin, U. Rudiger, C. Laubschat, *Phys. Rev. Lett.,* **100,** 107602 (2008).
27. M. Weser, Y. Rehder, K. Horn, M. Sicot, M. Fonin, A.B. Preobrajenski, E.N. Voloshina, E. Goering, Y.S. Dedkov, *Appl. Phys. Lett.,* **96,** 012504 (2010).
28. K.V. Emtsev, F. Speck, T. Seyller, L. Ley, *Phys. Rev. B.,* **77,** 155303 (2008).
29. P. Sutter, M.S. Hybertesen, J.T. Sadowski, E. Sutter, *Nanolett.,* **9,** 2654 (2009).
30. W. Meyer, K. Biedermann, M. Gubo, L. Hammer, K. Heinz, *J. Phys. : Cond. Matt.,* **20,** 265011 (2008).
31. C. Argile, G.E. Rhead, *Surf. Sci. Rep.,* **10,** 277 (1989).
32. D. Briggs and M.P. Seah, in *Practical Surface Analysis, 2nd Edition, Vol. 1, Auger and X-ray Photoelectron Spectroscopy,* D. Briggs and M. P. Seah, Editors, J. Wiley and Sons, New York (1990).
33. J. Park, W.C. Mitchel, L. Grazulis, H.E. Smith, K.G. Eyink, J.J. Boeckl, D.H. Tomich, S.D. Pacley, J.E. Hoelscher, *Adv. Mat.,* **22,** 4140 (2010).
34. E. Moreau, F.J. Ferrer, D. Vignaud, S. Godey, X. Wallart, *Phys. Stat. Solidi A,* **207,** 300 (2010).
35. P.A. Cox, *The Electronic Structure and Chemistry of Solids,* Oxford University Press, Oxford, 1987.
36. S.Y. Zhou, G.-. Gweon, A.V. Federov, P.N. First, W.A. de Heer, D.- H. Lee, F. Guinea, A.H. Castro Neto, A. Lanzara, *Nature Materials,* **6,** 770-776 (2007).
37. D. Cappus, C. Xu, D. Ehrlich, B. Dillmann, C.A. Ventrice Jr., K. Al Shamery, H. Kuhlenbeck, H.- J. Freund, *Chem. Phys.,* **177,** 533 (1993).
38. F. Rohr, K. Wirth, J. Libuda, D. Cappus, M. Baumer, H.- J.. Freund, *Surf. Sci.,* **315,** L977 (1994).
39. J. Goniakowski, C. Noguera, *Phys.Rev. B,* **66,** 085417 (2002).
40. P. Michetti, M. Cheli, G. Iannaccone, *Appl. Phys. Lett.,* **96,** 133508 (2010).
41. M. Cheli, P. Michetti, G. Iannaccone, *IEEE Trans. and Electron Dev.,* **57,** 1936 (2010).
42. X.Y. Chen, K.H. Wong, C.L. Mak, X.B. Yin, M. Wang, J.M. Liu, Z.G. Liu, *J. Appl. Phys.,* **91,** 5728 (2002).
43. H. Haugen, D. Huertas-Hernando, A. Brataas, *Phys. Rev. B,* **77,** 115406 (2008).

44. C. Mohapatra, J. Eckstein, *abstract for 2009 APS March meeting* http://meetings.aps.org/link/BAPS.2009.MAR.W26.12.

Figures and Captions

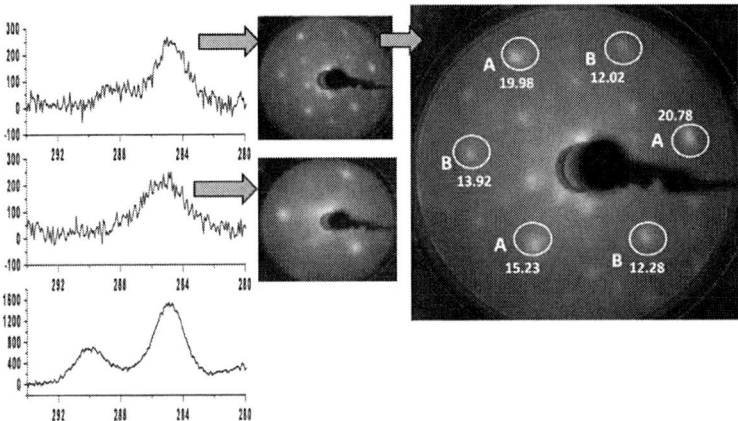

Figure 1. Evolution of C(1s) spectrum and corresponding LEED data for 1 ML graphene film on MgO(111) produced by annealing of adventitious carbon in UHV. C(1s) XPS and LEED spectra for formation of an ordered carbon monolayer on MgO(111): (a) XPS C(1s) spectrum of a multilayer adventitious carbon film observed immediately after insertion of the MgO crystal into UHV at room temperature (no LEED pattern is observed); (b) XPS C(1s) spectrum after annealing at 700 K in UHV in the presence of O_2; the XPS-derived average carbon layer thickness is1 ML; (c) the corresponding LEED pattern exhibits C3v symmetry; (d) XPS after anneal of the film in (b) to 1000 K in the presence of C_2H_4(5 $\times$ 10−7 Torr, 25 min); the XPS-derived C thickness remains 1 ML; (e) corresponding complex LEED pattern; (f) close- up of the patterning (e) with integrated, background-subtracted intensities (arbitrary units) for 'A' and 'B' spots (circled). The A spots have an average intensity of 18.7 $\pm$ 3, while the B spots have an average intensity of 12.9 $\pm$ 1. The uncertainties are the standard deviations. Other spots in the image are weaker and are attributed to multiple diffraction. The LEED patterns were acquired at 80 eV beam energy. The XPS spectra binding energies are referenced to a MgO lattice oxygen O(1s) binding energy of 530.0 eV. Data adopted from ref. [21], and used with permission.

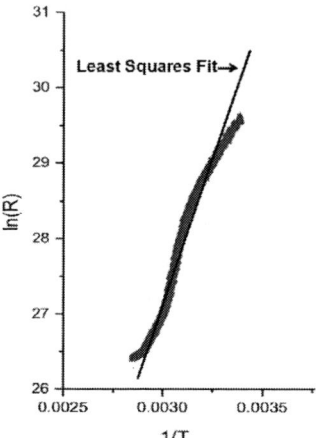

Figure 2. Plot of ln(Resistance) vs. reciprocal temperature for single layer of C(111) on MgO(111). Data shows semiconducting behavior with a charge carrier hopping activation energy of ~ 0.6 eV. Blue--actual data Black line--least squares fit.

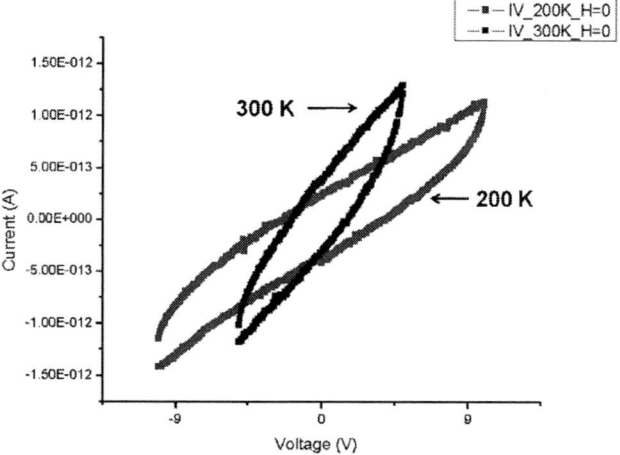

Figure 3. I-V curves, same sample as in Fig. 3. Note hysteresis

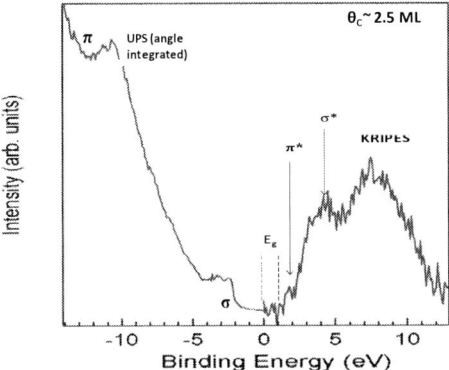

Figure 4. Angle-integrated valence band ultraviolet photoemission (UPS) and k-vector resolved inverse photoelectron spectroscopy (KRIPES) data for a 2.5 ML graphene film on MgO(111). The photoemission data correspond closely to spectra of high-quality films on transition metal substrates, but the data indicate a band gap of ~ 0.5 eV - 1 eV. The uncertainty is due to the limiting (0.4 eV) resolution of the inverse photoemission measurements, as well as final state effects in both spectra. Data adapted from ref. [20], and used with permission.

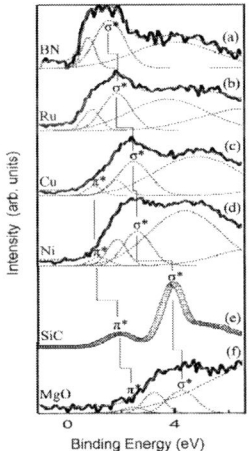

Figure 5. Inverse photoemission spectra of graphene on various substrates as labeled (background subtracted): (a)BN(0001)/Ru(0001); (b)Ru(0001); (c) Cu foil; (d) Ni foil; (e) SiC (adapted from ref. 12); (f)MgO(111). Adopted from Ref. [20] and used with permission.

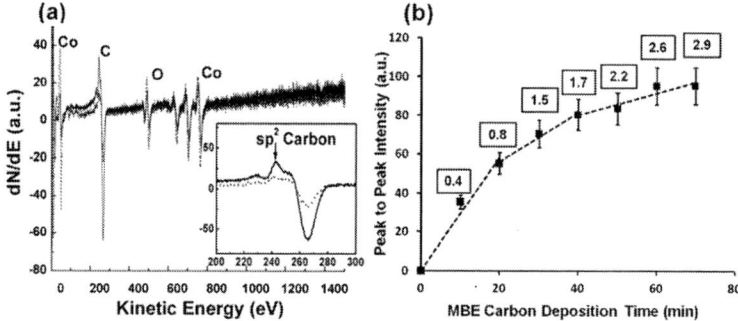

Figure 6. (a) Auger spectra acquired after deposition of 0.4 ML graphene (dashed line) and 3.0 ML graphene (solid line), inset—detail of C(KVV) spectral region showing characteristic sp² lineshape; (b) Growth curve showing evolution of Auger-derived average carbon overlayer thickness (number above each data point, ML) as a function of deposition time at 1000 K. Lines are least squares fits to the data. From Ref. [22] and used with permission.

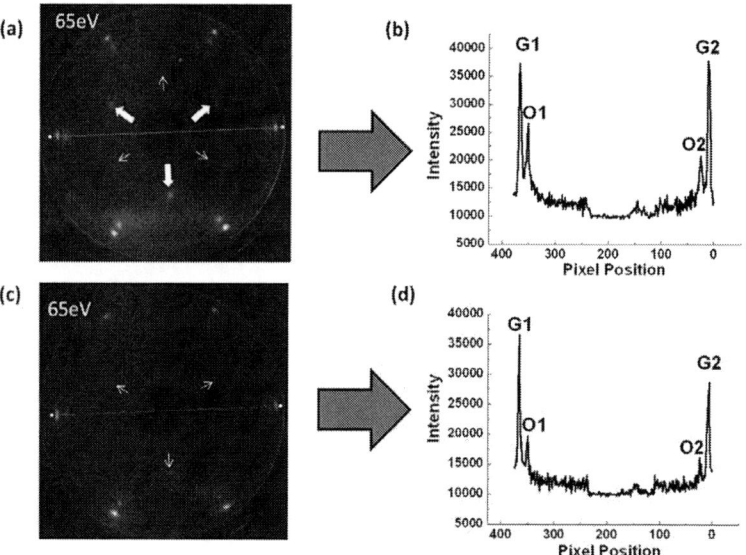

Figure 7. LEED and corresponding line scan data for (a,b) 0.4 ML graphene on $Co_3O_4(111)$, and (c,d) 3 ML graphene on $Co_3O_4(111)$. Arrows (a,c) mark diffraction spots associated with $Co_3O_4(111)$, as do inner spots in the outer ring of bifurcated features (e.g., O1,O2—b,d). Outer spots in the outer ring of bifurcated features (e.g., G1, G2—b,d) are graphene-related. LEED beam energy is 65 eV. From Ref. [22] and used with permission.

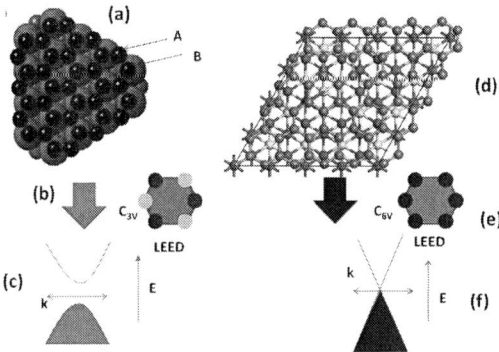

Figure 8. Graphene on a reconstructed, commensurate MgO(111) surface (left) vs. a bulk-terminated surface, forming an incommensurate interface (right), results in significantly different electronic properties. (a) Graphene (black spheres) on an O-terminated (O--red spheres; Mg--blue spheres) (111) surface, in which the O-O difference has been reconstructed to match the graphene lattice constant. Graphene A sites and B sites are now in chemically distinct chemical environments. This results in (b) a LEED pattern with three-fold symmetry, and also lifts the degeneracy of the HOMO and LUMO orbitals at the Dirac point, resulting in a band gap in (c). In contrast, on an unreconstructed surface as in (d) (graphene--gray spheres; O--red spheres; Mg--green spheres) both A and B sites in the graphene lattice are in an ensemble of chemical environments, resulting in (e) C_{6v} symmetry in the LEED pattern and (f) retention of HOMO/LUMO degeneracy at the Dirac point (no band gap).

Aberration Corrected Microscopy of CVD Graphene and Spectroscopic Ellipsometry of Epitaxial Graphene and CVD Graphene for Comparison of the Dielectric Function

Florence Nelson*, Dhiraj Prasad Sinha, Everett Comfort, Ji Ung
Lee, Alain C. Diebold
*College of Nanoscale Science and Engineering, SUNY at
Albany, NY*
*Corresponding Author

Andreas Sandin, Daniel B. Dougherty,
Jack E. Rowe
Department of Physics, NCSU, Raleigh, NC

Graphene's importance in post-CMOS device research drives the need for advancing a variety of metrology methods. Chemical Vapor Deposition (CVD) on metallic foils and the thermal decomposition of SiC have become two of the dominant fabrication methods of large area graphene due to their industrial scalability. Electrical and optical characterization of CVD graphene requires transfer of the graphene film to secondary substrates (i.e. SiO_2/Si), but is conducive to TEM imaging due to the fact that the film can be etched from the growth foil and directly transferred to support grids. Epi graphene does not require film transfer for optical or electrical characterization and can control layer number by selection of process parameters as well as Si-face vs. C-face growth. The work will present STEM imaging of CVD graphene performed at 60 kV with atomic resolution achieved through aberration correction. Single-crystal areas are identified as well as point defect structures. Spectroscopic Ellipsometry (SE) is used for the optical characterization of the CVD prepared graphene as well as the epitaxial graphene for a comparison of the dielectric function. Previous SE work on CVD graphene has shown the validity of an isotropic model assumption for a single layer of graphene due to the insensitivity of an SE measurement to anisotropy for sub nm thicknesses (3.35 Angstroms) (1). However a comparative simulation study of isotropic vs. anisotropic modeling showed the necessity of an anisotropic model for films greater than two layers. We therefore present an anisotropically modeled stack for the 3-4 layer epitaxial graphene grown on 6H SiC, as well as the incorporation of a "buffer layer" present between the graphene and SiC substrate. We likewise observe spectral differences in the associated dielectric functions which we assume to the different bonding schemes present in the two regions.

Introduction

Graphene growth methods have increased in scalability since the initial "scotch tape" method of exfoliating single-layer flakes. The two growth methods usually sited for their large area graphene sheets are chemical vapor deposition (CVD) onto metallic foils, and thermal annealing of SiC. For the present work, graphene from both growth techniques was measured with Spectroscopic Ellipsometry (SE) for the study/comparison of the dielectric function/complex refractive index (CRI). Additionally, CVD graphene was etched from the growth foil and transferred directly to support grids for imaging with Scanning Transmission Electron Microscopy (STEM). Aberration-correction at 60 kV allowed for atomic resolution and observation of defect structures within the lattice. We find qualitative agreement between the two growth methods for the main spectral features for graphene SE (i.e. an exciton dominated absorption peak in the UV and a fairly constant absorbance from the visible to IR regions), with some differences which may be due to the growth method and/or layer number. Aberration-corrected STEM imaging of the graphene lattice shows atomic resolution and the observation of defect structures within one domain orientation of the CVD-grown films.

A main difference between the SE analysis for CVD graphene transferred to a secondary substrate vs. epitaxial graphene grown on SiC is the presence of a buffer layer between the film and the substrate in the latter sample. This buffer layer, thought to consist of a combination of sp^2 and sp^3 bonded carbon, must be taken into consideration if accurate thickness determinations are to be made using optical in-line thickness metrology methods such as SE. This buffer layer has been investigated by several groups with regards to thickness as well as sp^2/sp^3 bonding ratio. We compare our results to previous work that has been done on composite carbon sp^2/sp^3 bonded films using an effective medium approximation, as well as XPS studies of epitaxial graphene grown on the Si-face of SiC in order to approximate the percentage of sp^2 bonding in the buffer layer. The epitaxial graphene in this work was grown on vicinal surfaces on 6H-SiC (0001) in order to provide a higher density of step edges where the Si sublimation occurs.

Experimental

CVD Graphene Growth

CVD graphene was grown by a combined flow of hydrogen and methane onto Cu foils and transferred to glass slides as described elsewhere (1). Decomposition of the methane onto the Cu results from a surface catalyzed reaction that leaves predominantly monolayer graphene due to carbon's insolubility in Cu (2). It has been shown that control of process parameters such as temperature and precursor flow rate can produce smaller or larger grains as a results of more or less nucleation sites, respectively (3). The grain size of the films may be found from Dark-Field TEM (DF-TEM) (4). This method takes advantage of the polycrystalline diffraction pattern generated by a film with multiple domain orientations. An objective aperture is used to select one or multiple reflections in the back focal plane in order to only image domains of a certain orientation (4). The presence of monolayer graphene was verified using Raman spectroscopy.

Epitaxial Graphene Growth

Epitaxial graphene was grown by the thermal annealing of SiC. The growth results from the sublimation of Si from the step edges of the SiC terraces (5). Growth rates differ based on temperature, as well as whether the C-face or Si-face is considered. 6H-SiC Si(0001) substrates that had been miscut by 3.5° in the [11-20] direction were used to provide a higher density of these step edges in order to facilitate the Si sublimation. Substrates were cleaned by CMP and HF etching and were annealed at ~900°C prior to growth in order to remove the native oxide. The growth was done in UHV at ~1400°C for ~3 minutes. Vicinal 6H-SiC substrates were purchased from CREE.

Figure 1 is an STM topology of the graphenized sample showing that step-bunching has resulted from the growth process. This occurs in order to minimize the surface energy of the SiC bilayer steps. The majority of steps on the vicinal 6H-substrates before the growth are one SiC bilayer high. However, the more energetically favorable step height is three SiC bilayers, or ½ of a 6H-SiC unit cell (6). Therefore the three-bilayer step heights overtake the single bilayers during the growth process while allowing a uniform graphene film to form on top of them. Interestingly, it was observed that a higher growth temperature was required for these off-axis samples in comparison to on-axis samples, possibly due to a more stable surface reconstruction prior to the graphenization.

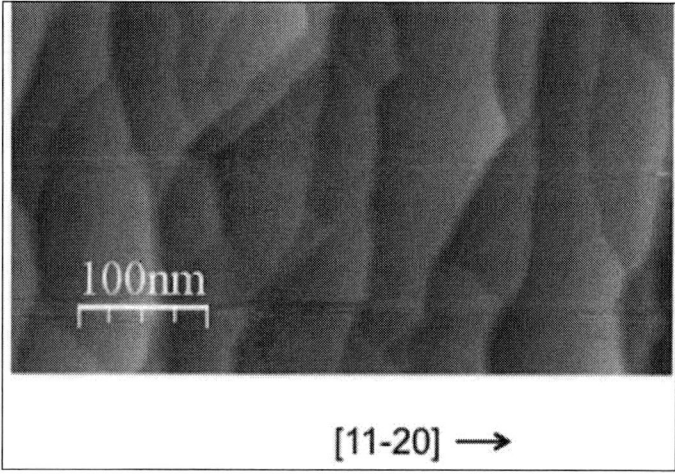

Figure 1. STM image of graphenized SiC showing step bunching. Prior to graphene growth the average step width was 5.5 nm with a height of 0.34 nm while afterwards the average height increased by 3x to ~1 nm and the width varied from ~5 to more than 80 nm.

Results and Discussion

STEM Imaging of CVD Graphene

CVD graphene grown on atomic foils is conducive to TEM sample preparation since the substrate may be easily etched away in a solution of FeCl. A process similar to that of Regan *et al.* (7) was used in which a TEM grid is placed on the graphene while it is still on the Cu foil. A droplet of IPA is placed on the grid so that its evaporation binds the grid to the graphene. The foil with the grid on top is then placed in a bath of FeCl. After the copper is etched away, the grid with graphene is rinsed in DI water. This fairly simple process leaves well-adhered graphene that may be subsequently annealed in order to remove adsorbates prior to imaging

CVD graphene samples were imaged with Scanning Transmission Electron Microscopy (STEM) at 60 kV using a Nion UltraSTEM. Aberration-corrective optics allow for atomic resolution of the graphene lattice, as well as observation of defect structures (8,9). STEM allows for direct observation of the atomic positions as compared to High Resolution TEM (HR-TEM), or phase contrast images, which are interference patterns of the transmitted beam with those diffracted at different defocus values. Figure 2 shows the atomically-resolved graphene lattice of CVD-grown graphene with a vacancy defect circled in red. The sample was annealed for 10 hours at 160°C prior to imaging.

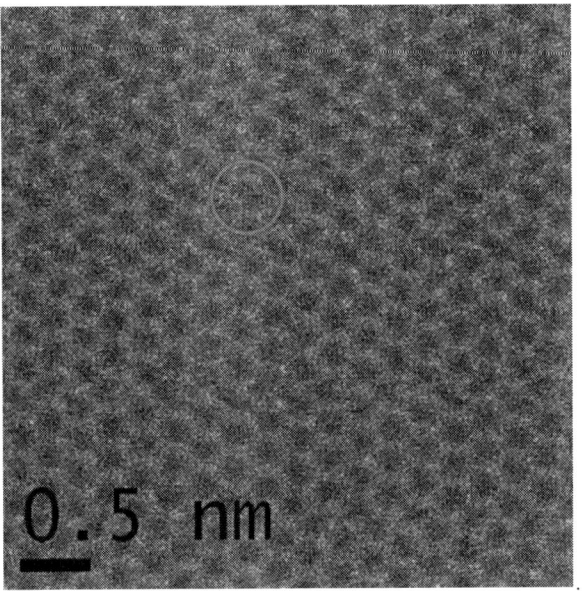

Figure 2. Atomically-resolved image of CVD graphene with circled vacancy.

Ellipsometry of CVD and Epitaxial Graphene

SE allows for measurement of a thin film's thickness or dielectric function (CRI) from the change of the polarization state of light after reflection from a surface. The two measured quantities, ψ (the real part reflection coefficient ratio) and Δ (the imaginary part of the reflection coefficient ratio), may be analyzed using oscillator models to represent absorptions at particular energies (imaginary part of the dielectric function, ε_2). The Kramer's-Kronig transform (Equation 1) is then used to find the real part of the dielectric function.

$$\varepsilon_1(E) - 1 = \frac{2}{\pi} P \int_0^\infty \frac{E' \varepsilon_2(E')dE'}{E'^2 - E^2}$$

Eq. 1

Both the CVD and epitaxial graphene samples were measured in air using a J. A. Woollam dual-rotating compensator ellipsometer (RC2^{TM}) from ~0.7-5 eV. While adsorbates and contamination are ubiquitous on graphene samples exposed to the environment, several groups have shown agreement for graphene's dielectric function/optical absorbance when the samples have been measured using SE as well as transmission/reflectance spectroscopy (1, 10, 11, 12, 13). We therefore compare the dielectric function of both the CVD and epitaxial samples relatively, assuming similar effects from environmental contaminants.

Previous work (1) has demonstrated the use of an isotropic model for monolayer graphene due to the limited path length of light through the material causing an insensitivity to anisotropy (14). Therefore the CVD graphene was modeled as an isotropic layer with a predetermined thickness of 3.35Å. However, the isotropic assumption begins to break down when a graphene thickness of ~2-3 layers is reached (14). Auger analysis of the epitaxial graphene samples estimated a film thickness of ~ 4 layers. Therefore an anisotropic model was used for the analysis of the epitaxial graphene, with the extraordinary (z-response) refractive index modeled by a Cauchy dispersion. The experimental and modeled fits of the epitaxial sample's ψ and Δ spectra are shown in Figure 3.

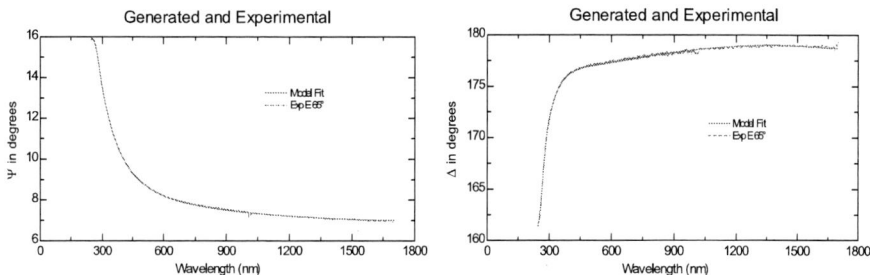

Figure 3. Experimental and modeled spectra of ψ and Δ for epitaxial graphene.

The dielectric functions of both the CVD and epitaxial graphene are shown in Figure 4. Both agree qualitatively in terms of a UV absorbance peak at ~ 4.5 eV, the so-called van Hove peak, and an in-plane response similar to that of bulk graphite. Our work has observed the intensity of the van Hove peak to be quite sample dependent with respect to choice of substrate, graphene domain size, etc. in the measurement of single-layer graphene. This is apparent in the difference in peak intensities for the CVD and epitaxial ε_2 values in Figure 4. However, the epitaxial sample also shows a slight blue-shift in the position of the van Hove peak with respect to the CVD-grown sample. Since the epitaxial film consists of more layers (~4) than that produced from CVD, one possibility is greater screening in the epiaxial graphene layers, which would reduce the excitonic effects that red-shift the van Hove peak to ~4.5 eV in monolayer graphene (15).

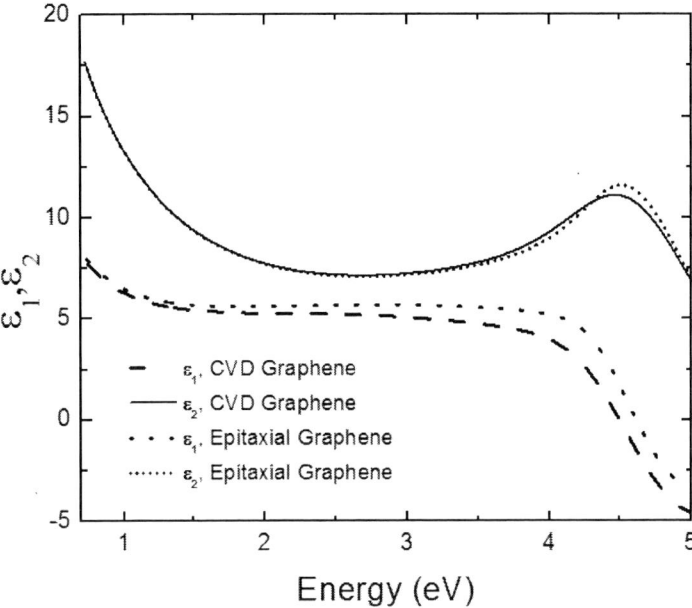

Figure 4. Dielectric functions of CVD graphene and epitaxial graphene prepared from the thermal decomposition of SiC.

It is well know that graphene grown on the Si-face of SiC rests on a buffer layer or interface layer between the substrate and pure graphene layers (16). This layer is thought to consist of a combination of sp^2 and sp^3 bonded carbon in order to accommodate the SiC substrate. This layer is of interest from an optical metrology standpoint as accurate thickness determinations will be required for device applications. We therefore extracted the dielectric function of this region using a fixed layer thickness of 2.5 Å (SiC bilayer), as shown in Figure 5. In comparison to the van Hove peak for the graphene layers, the buffer layer peak has red-shifted to a lower energy at ~4.0 eV which we attribute to the sp^3 bonded carbon. Comparison to previous work (17) on composite films of sp^2 and sp^3 bonded carbon suggest this layer is more heavily sp^2 bonded, which agrees with ARPES studies reported for graphene films grown on the SiC(0001) (18).

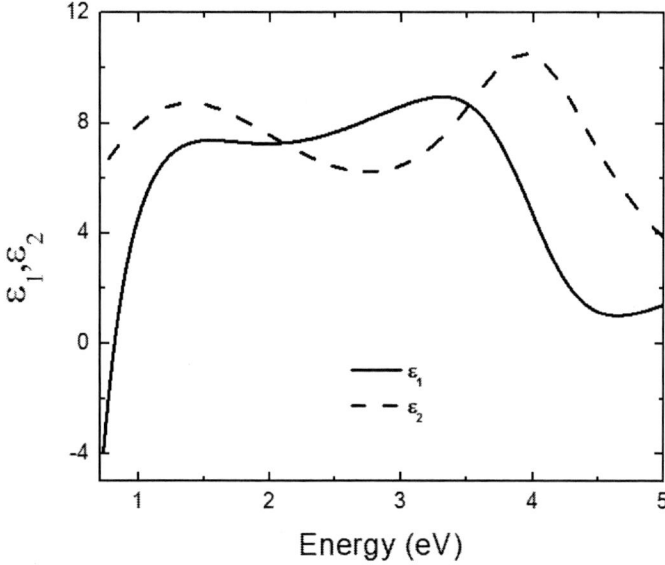

Figure 5. Dielectric function of buffer layer below epitaxial graphene.

Conclusions

Both CVD graphene and epitaxial graphene on SiC have been measured with SE for comparison of the dielectric function. A slight difference in the van Hove peak energy was observed possibly due to the differences in layer number and/or growth method. The buffer layer dielectric function of the latter growth method has also been extracted and showed differences when compared to the purely sp^2 bonded graphene layers. Aberration corrected STEM imaging allowed for atomic resolution of the CVD graphene lattice and defect structures.

Acknowledgements

The author's gratefully acknowledge Juan-Carlos Idrobo at Oak Ridge National Labs (ORNL) for his assistance in aberration-corrected STEM imaging of CVD graphene and Dave Aspnes at NCSU for very useful discussion. Funding for this work at the College of Nanoscale Science and Engineering was provided by INDEX/NRI. NCSU support was provided by NSF through the Center for Molecular Spintronics. This research was supported in part by ORNL's

Shared Research Equipment (SHaRE) User Facility, which is sponsored by the Office of Basic Energy Sciences, U.S. Department of Energy.

References

1) F. Nelson, *et al., App. Phys. Lett.,* **97**, 253110 (2010).
2) X. Li *et al., Science*, **324**, 1312 (2009).
3) X. Li *et al., Nano Lett.,* **9**, 4268 (2009).
4) P. Y. Huang *et al., Nature,* **469**, 389 (2011).
5) M. Hupalo *et al., Phys. Rev. B,* **80**, 041401 (2009).
6) T. Kimoto *et al., J. Appl. Phys.,* **81** , 8 (1997).
7) W. Regan, *et al., App. Phys. Lett.,* **96**, 113102 (2010).
8) J. C. Meyer *et al., Nano Lett.,* **8**, 3582 (2008).
9) F. Banhart *et al., ACS Nano*, **5**, 26 (2011).
10) V. G. Kravets *et al., Phys. Rev. B,* **81**, 155413 (2010).
11) J. W. Weber *et al., Appl. Phys. Lett.,* **97**, 091904 (2010)
12) K. F. Mak *et al., Phys. Rev. Lett.,* **101**, 196405 (2008)
13) D. H. Chae *et al. Nano Lett.,* **11**, 1379 (2011);
14) M. Losurdo and K. Hingerl, Ed. Ellipsometry at the Nanoscale. Springer. In prep.
15) L. Yang et al., Phys. Rev. Lett., **103**, 186802 (2009).
16) W. A. de Heer *et al., Sol. State Comm.,* **143**, 92 (2007).
17) S. Logothetidis, *Diam. Rel. Mat.* **12**, 141 (2003).
18) C. Riedl. Epitaxial Graphene on Silicon Carbide Surfaces: Growth, Characterization, Doping and Hydrogen Intercalation. *Thesis* (2010).

Modeling the Growth of SWNTs and Graphene on the Atomic Scale

E. C. Neyts[a], A. Bogaerts[a]

[a] University of Antwerp, Department of Chemistry, Research Group PLASMANT
Universiteitsplein 1, 2610 Wilrijk-Antwerp, Belgium

The possibility of application of nanomaterials is determined by our ability to control the properties of the materials, which are ultimately determined by their structure and hence their growth processes. We employ hybrid molecular dynamics / Monte Carlo (MD/MC) simulations to explore the growth of SWNTs and graphene on nickel as a catalyst, with the specific goal of unraveling the growth mechanisms. While the general observations are in agreement with the literature, we find a number of interesting phenomena to be operative which are crucial for the growth, and which are not accessible by MD simulations alone due to the associated time scale. Specifically, we observe metal mediated healing and restructuring processes to take place, reorganizing the carbon network during the initial nucleation step. In the case of carbon nanotube growth, this leads to the growth of tubes with a determinable chirality. In the case of graphene formation, we find that graphene is only formed at temperatures above 700 K. These results are of importance for understanding the growth mechanisms of these carbon nanomaterials on the fundamental level.

Introduction

Carbon nanotubes (CNTs) continue to attract much attention thanks to their extraordinary properties. For instance, CNTs may be used as interconnects in silicon IC manufacturing as their current carrying capacity exceeds 10^7 A.cm^{-2}. Also, the very high thermal conductivity of CNTs has potential for their use in dissipating heat from the chips [1]. Other possible applications include CNT-based composites, hydrogen storage, and their use as electrochemical devices, field emission devices, gas sensors and probes. However, their applicability is currently limited due to their production cost, polydispersity in nanotube type and assembly methods [2]. Clearly, a more thorough understanding of their growth mechanisms would help to overcome these barriers.

Graphene is currently attracting even more interest, mainly because of its unusual electronic properties. From the application point of view, interesting graphene properties include quantum electronic transport, a tunable band gap, extremely high mobility or electromechanical modulation [3-6].

Both CNTs and graphene can be produced by catalytic chemical vapor deposition (CCVD). CCVD operates by decomposition of hydrocarbons (such as CH_4, C_2H_2, etc.) on metal catalysts (Ni, Fe, Co, FeMo, etc.) [7-9]. The role of the catalyst is to actively decompose the hydrocarbon molecules into atomic carbon.

In the case of CNT growth, various models have been proposed, of which the vapor-liquid-solid (VLS) model is the most widely accepted model for typical production conditions. In the VLS model, the catalyst particle is in the liquid state, which allows rapid diffusion of carbon atoms throughout the particle [10]. After supersaturation, the carbon atoms start to segregate at the surface, and form a solid crystalline CNT.

In the case of graphene, the growth process in CCVD growth was experimentally determined to be surface segregation, followed by precipitation during the cooling step.

Despite numerous studies, however, the mechanisms operative on the atomic scale, are still far from completely understood. Experimentally, for example, it is extremely difficult to accurately measure the temperature of the metal nanoparticles during the growth process, or to follow the dynamic growth process on the atomic level. While these issues could in principle be resolved by computer simulations, the methodologies currently available are prohibitively time expensive to simulate the growth process at realistic growth rates. Indeed, processes such as surface diffusion and bulk diffusion, which are of paramount importance in both CNT and graphene CCVD growth, occur on time scales well beyond the MD time limit of ca. 100 ns. The difference in time scales between simulation and experiment makes a direct comparison difficult.

Therefore, in this contribution, we focus on the growth mechanisms of both CNTs and graphene, with specific emaphasis on the effect of taking into account long time scale events.

Methodology

We employ atomic scale simulations to unravel the mechanistic details of the growth process. These simulations consist of a hybrid molecular dynamics / Monte Carlo (MD/MC) approach, in which the forces and between the atoms are derived from the ReaxFF interactomic potential [11]. ReaxFF is based on the bond energy / bond order and bond order / bond length relationship. The bond order is summed over sigma, pi and pi-pi contributions, and subsequently energy penalties are applied to enforce the correct bond order. The total energy is calculated as a sum over various partial energy terms, related to lone pairs, over- and undercoordination, conjugation, valence and torsion angles, hydrogen bonding, Coulomb interactions and van der Waals interactions. ReaxFF is therefore capable to describe the entire range of chemical bonding, from covalent to ionic and anything in between.

In the hybrid MD/MC simulation, MD and MC stages are alternating, each providing the input structure for the next stage [12-14]. Each MD stage runs for a preset time of several picoseconds, using a timestep of 0.25 fs employing a velocity Verlet integration scheme. The temperature is controlled using the Berendsen thermostat with coupling constants of 100 fs. In the case of CNT growth, the temperature was set to 1000 K. In the case of graphene growth, the temperatures investigated are 300 K, 500 K, 700 K, 900 K, 1100 K, and 1273 K.

We use a small floating Ni-nanocluster containing several tens of atoms in the case of CNT growth. In the case of graphene growth, the substrate is a Ni(100) structure with a thickness of about 6 Å, corresponding to 4 atomic layers of 50 atoms each, covering a total surface area of about 317 $Å^2$. The lowest atomic layer is kept fixed. Periodic boundaries are applied in the lateral directions to simulate a semi-infinite surface.

Results

In the following sections, we describe our results on the growth of SWNTs and graphene.

SWNT growth

In Figure 1, a respresentative structure of a SWNT as grown in the (a) the hybrid MD/MC simulations, and (b) pure MD simulations is shown.

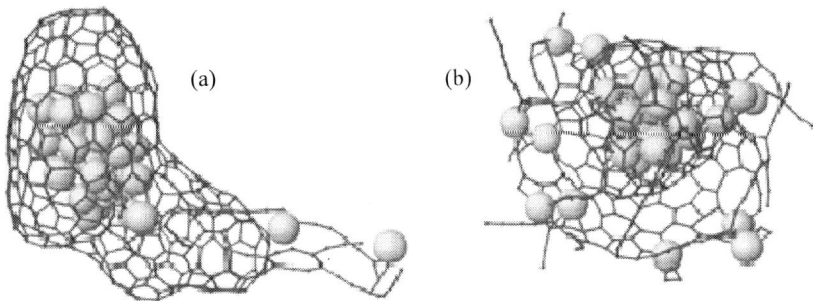

Figure 1. Structures grown in the (a) hybrid MD/MC simulations, resulting in SWNT growth, and (b) pure MD simulations, resulting in the formation of amorphous carbon. The blue balls represent nickel atoms; the red and green atoms represent three- and two-coordinated carbon atoms, respectively.

From Figure 1, it is clear that applying the hybrid methodology, i.e., MD/MC, is required in order to obtain a well-structured SWNT cap. Using this methodology, the formation mechanism of the SWNT cap is determined to be as follows. Initially, the carbon atoms impinge on the metal surface and dissolve into the bulk of the particle. The atoms diffuse through the bulk, and occasionally reappear at the surface. When sufficient carbon atoms are dissolved in the particle, new impinging carbon atoms start to form bonds at the surface. Initially, a number of dimers and trimers are formed, located randomly over the surface. Dimers and trimers are also found to a limited extend in the subsurface region of the particle, but not in the bulk [12]. When more carbons are added to the system, longer chains are formed, which fold with the formation of pentagons, hexagons and heptagons. Thus, these initial rings are also scattered more or less randomly over the surface, rather than being connected to each other. This is a direct consequence of the inherently random nature of the addition of the carbon atoms to the metal particle. However, when more rings are being formed, these rings eventually coalesce, with the formation of a small number of initial graphene patches. Finally, also these patches coalesce, with the formation of a relatively large graphene layer which detaches from the metal surface, thereby forming the SWNT cap [13].

When applying the same ReaxFF force field in the pure MD simulations, i.e. without the additional MC step, we do not observe the formation of a SWNT cap, but rather the formation of amorphous carbon. This is shown in Figure 1b. Indeed, it is found that the most important reason for the formaton of well structured graphene patches is the occurence of a metal mediated healing and network restructuring mechanism. In this mechanism, pentagons, heptagons, and other topological defects are healed into hexagons. Most probably, this is related to the lowering of the activation barrier for these processes, similar to the lowering of the activation barrier for healing of the Stone-Wales defect [14]. This mechanism, as well as the accompaning restructuring of the carbon network, seems to take place on a time scale which is beyond the reach of pure MD alone. Therefore, application of the MC method, in combination with the MD method, allows to take into account long time scale events, which finally results in the formation of a SWNT cap with (12,4) chirality [13]. It should be realized, however, that in these simulations, we have no information on the duration of these events or on the simulated time scale.

Graphene growth

A pictorial view of the final state of an MD/MC simulation for graphene growth at a temperature of 900 K is presented in Figure 1. The sequence of observed events can be described as follows. At all investigated temperatures, the impinging carbon atoms initially adsorb on the surface, where they diffuse either from surface site to surface site, or possibly to subsurface sites. As the concentration of carbon at the surface increases (due to the continued addition of carbon atoms from the gas phase), also the occupation of the subsurface sites increases, along with the appearance of the first dimers and trimers at the surface. These surface dimers and trimers are observed to be highly mobile, especially at higher temperature. The incorporation of carbon atoms into interstitial sites distorts the Ni-crystal structure, which results in a semi-amorphous top layer. Subsequently, as the number of carbon atoms in/on the surface keeps on increasing, somewhat longer and possibly branched chains up to 5 or 6 carbon atoms start to appear. These chains virtually always adopt valence angles of about 120°. Hence, some of these chains fold upon themselves with the formation of the first rings. These first rings serve as the initial nuclei from which a graphene sheet can grow in a later stage. These structures, containing 4 to 7 carbon atoms, are also observed to diffuse on the surface. Note that the initial rings only appear at the surface, not in the bulk of the crystal. However, we have observed the formation of a few dimers in the subsurface region of the film, especially at the higher temperatures. This corresponds to our observations during SWNT growth. A further increase in the number of carbon atoms allows these first rings and surface chains to grow until they meet. Continuing this process of addition of carbon atoms, the growing of the existing carbon structures at the surface and their coalescence, as well as the addition of carbon atoms originating from subsurface sites to these surface structures, finally leads to the formation of an almost complete graphene-like layer at all temperatures on top of a carbon saturated Ni-crystal surface.

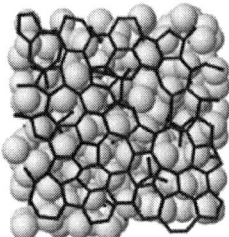 

Figure 1 – Final configurations of graphene grown on a Ni surface at 900 K, obtained from the hybrid MD/UFMC simulations, in top view and side view. The blue balls represent Ni atoms; the black lines are the bonds between the gold-colored carbon atoms.

Although this general picture is observed at all temperatures, the details of the formation process are different, resulting in different final structures at the various temperatures investigated. Based on the resulting carbon layers, we can distinguish two temperature regimes. In the lower temperature range (300 K – 500 K) a very defective layer is obtained, whereas at medium and high temperatures (700 K – 1273 K), the quality of the layer appears to be much higher.

At the lower temperatures (300 K – 500 K), the first dimers and trimers are formed by the impingement of carbon atoms in the vicinity of already incorporated carbon atoms. As the solubility of carbon in nickel at low temperature is limited, the surface and subsurface area quickly become saturated, and newly added carbon atoms can only be deposited at the surface, thereby increasing the extent of the already existing rings and chains. However, as the mobility of the carbon atoms at low temperature is very limited, the formation of a carbon network at the surface is essentially determined by the consecutive impingement locations of the carbon atoms from the gas phase. As a result, very defective layers are formed. Furthermore, the low temperature inhibits the healing of defects.

At higher temperatures (700 K – 1273 K), the first dimers and trimers are formed not only by the impingement of carbon atoms close to other carbon atoms, but also because of the increased mobility of the carbon atoms. In this case, the solubility of carbon in the nickel film is limited by the extent (or thickness) of the film in our simulations. Hence, also at higher temperature, the point of saturation is quickly reached, as there is no bulk crystal that can act as a sink to which the carbon atoms can diffuse. Newly added carbon atoms will therefore also be deposited at the surface at higher temperature. In this case, however, the carbon atoms (as well as the dimers and trimers) can easily diffuse over the surface in order to find the energetically most stable location, i.e., their incorporation in a hexagonal network. Furthermore, the higher temperature aids in the metal-mediated healing of defects, as we have also observed earlier in the growth of single-walled carbon nanotubes on Ni-clusters (see above) [13]. Both processes lead to an increase in the observed number of hexagons as a function of temperature.

Conclusions

We have applied hybrid MD/MC simulations to the catalyzed thermal growth of SWNT and graphene on nickel nanoparticles and surfaces, respectively. We found that applying the MC step is necessary to take into account longer time scale events which are beyond the reach of pure MD, which results in the formation of well structured configurations. The growth mechanism of the SWNT and graphene layer is found to be similar, consisting of a number of distinct steps: dissolution of carbon in the metal; (super)saturation and the formations of first rings; concatenation of these rings forming small graphitic patches; coalescence of these patches; and finally detachment and growth of the cap in the case of SWNT growth and growth of the graphene layer in the case of graphene growth.

References

1. M. Meyyappan, L. Delzeit, A. Cassell, and D. Hash, *Plasma Sources Sci. Technol.* **12**, 205 (2003).
2. R. H. Baughman, A. A. Zahkidov, and W. A. de Heer, *Science* **297**, 787 (2002).
3. Y. Zhang, Y. W. Tan, H. L. Stormer, and P. Kim, *Nature* **438**, 201 (2005).
4. M. Y. Han, B. Oezyilmaz, Y. Zhang, and P. Kim, *Phys. Rev. Lett.* **98**, 206805 (2007).
5. K. I. Bolotin, K. J. Sikes, Z. Jiang, M. Klima, G. Fudenberg, J. Hone, P. Kim, and H. L. Stormer, *Solid State Commun.* **146**, 351 (2008).
6. J. S. Bunch, A. M. van der Zande, S. S. Verbridge, I. W. Frank, D. M. Tanenbaum, J. M. Parpia, H. G. Craighead, and P. L. McEuen, *Science* **315**, 490 (2008).

7. J. Kong, H. T. Soh, A. M. Cassell, C. F. Quate, and H. Dai, *Nature* **395**, 878 (1998).
8. D. B. Geohegan, A. A. Puretzky, I. N. Ivanov, S. Jesse, G. Eres, and J. Y. Howe, *Appl. Phys. Lett.* **93**, 1851, (2003).
9. A. R. Harutyunyan, B. K. Pradhan, U. J. Kim, G. Chen, and P. C. Eklund, *Nano Lett.* **2**, 525 (2002).
10. R. S. Wagner, W. C. Ellis, *Appl. Phys. Lett.* **4**, 89 (1964).
11. A. C. T. van Duin, S. Dasgupta, F. Lorant, W. A. Goddard III, *J. Phys. Chem. A* **105**, 9396 (2001).
12. E. C. Neyts, A. C. T. van Duin, A. Bogaerts, *J. Am. Chem. Soc.* **134**, 1256 (2012).
13. E. C. Neyts, Y. Shibuta, A. C. T. van Duin, A. Bogaerts, ACS Nano **4**, 6665 (2010).
14. E. C. Neyts, A. C. T. van Duin, A. Bogaerts, *J. Am. Chem. Soc.* **133**, 17225 (2011).

Large Area Mapping of Graphene Grain Structure and Orientation

H.C. Floresca[a], D. Hinojos [a], N. Lu [a], J. Chan [a], L. Colombo [b], R. M. Wallace [a], J. Wang [a], J. Kim [a] and M. J. Kim [a]

[a] Department of Materials Science and Engineering, The University of Texas at Dallas, Richardson, Texas 75080, USA
[b] Texas Instruments, Dallas, Texas 75243, USA

Chemical vapor deposited (CVD) graphene grows as a polycrystalline sheet with grains that are oriented in many directions. There is a push to increase the size of these grains until a large single-crystal graphene sheet can be achieved. With the increase in grain size comes the need to be able to determine the lattice orientation of larger areas. Here we present a method using a simple well-known transmission electron microscopy (TEM) technique that allows us to map the orientation of a 50μm by 50μm area of graphene. This method created the largest graphene orientation map to date and has the ability to scale with the increasing grain sizes. From the map, statistics and interpretations were drawn, giving details about the graphene grown through the CVD recipe.

CVD Graphene Mapping

A perfect graphene sheet is a two-dimensional structure made of carbon atoms in a one atom thick hexagonal lattice. CVD graphene is far from perfect since during its growth on a copper substrate, it begins as individual islands containing multiple lattice orientations. As these islands grow to completely cover the copper a single sheet is formed locking in the different orientations as grains. Studies to map these grains have been done using LEED and TEM, but may soon be useless when the size of grains increases larger than their field of view. In our study, we show a simple method to identify and analyze the orientations of these lattices that make up a sheet of graphene and has the ability to scale to larger sizes with ease.

Sample Preparation and Method

The CVD graphene used in this experiment was grown on a copper foil until graphene completely covered its surface (1). The foil was then dissolved away to transfer the graphene onto a silicon TEM frame that contained a 50μm square window. This window was covered by a SiN film that is used as a support for the transferred graphene, allowing it to lay flat and not sag over the open area.

1μm selected area diffraction (SAD) patterns were taken throughout the whole sample with a 1μm pitch, resulting in over 2,500 patterns. Each pattern had their position on the sample recorded for mapping purposes. Within each diffraction pattern is a six-fold spot image caused by the electron diffraction off of the graphene lattice. This image is sensitive to the rotation of the lattice which is utilized in determining the rotation of the originating lattice. The rotation of the diffraction pattern corresponds to the same rotation in the actual lattice. Therefore, measuring the pattern rotation reveals the orientation of the original crystal. Each pattern is measured for this rotation and recorded. This information is then utilized by a customized program to analyze and plot in an understandable manner. The output of the program is overlaid on a TEM image seen in figure 1a. A legend is included indicating the colors assigned to the measured lattice rotation. In the legend, the reference angle of 0° is shown to be at the 12 o' clock position. The angles increase in a counterclockwise direction until 60°. Due to six-fold symmetry, the lattice can only rotate by 59° until it will repeat itself at 0°. Areas of black pixels contain diffraction patterns that did not form hexagonal patterns which may be due to an amorphous area, torn graphene or pinholes in the sample.

The computer program was also able to take the information and represent them in a chart form. An example is seen in figure 2a. The program counted the number of times each rotation measurement was found. This allowed us to determine which rotations made up most of the sample and which ones were grown the least. Since every diffraction pattern sampled a micrometer area, the count closely indicates the area of coverage for that certain orientation. Thus, the combined graphene lattices which are rotated by 54° covers about 86μm of the sample window.

Results and Discussion

To understand the shapes of the graphene grains, the raw data was interpreted to give better estimates of where grain boundaries lie. The raw data interpretation can be seen in figure 1b. It has been reported that grains originate from graphene island nucleation sites and radiate outwards (2). Using the shapes of the grains in the interpreted data, the grown islands were estimated and identified by black bold lines in the figure.

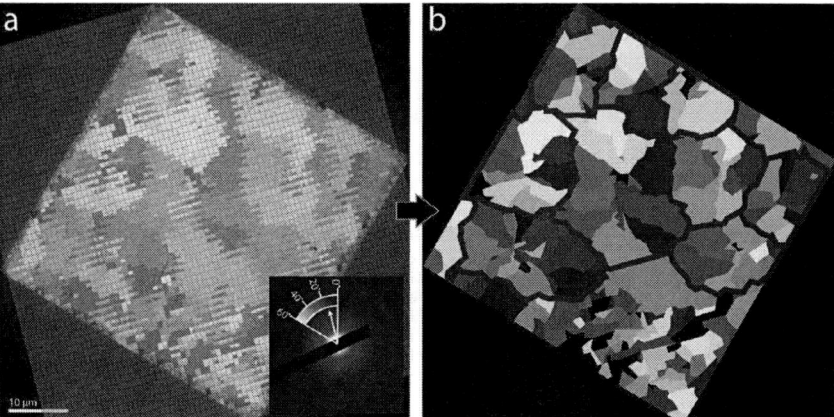

Figure 1. (a) The overlaid program output on a TEM image. The inset legend shows how the pixels in the output are designated a color based on the diffraction pattern image. (b) The interpreted data showing the location of the islands and the merged grain boundaries.

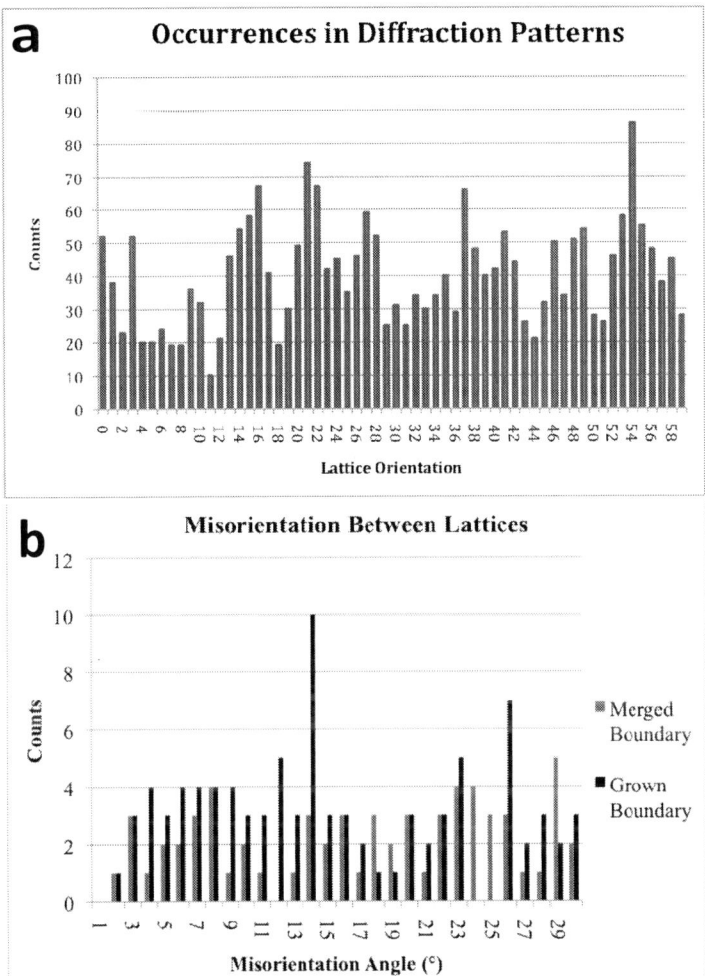

Figure 2. (a) A chart of the diffraction pattern information showing a higher coverage of graphene that is oriented in the 54° direction. (b) The measured grain boundary misorientations between merged boundaries and grown boundaries. Grown grain boundaries have a preferential misorientation at 14°.

Measurements between grains were conducted to extract the misorientation angles between grains. Two types of grain boundaries were measured and recorded. One type is the grown grain boundary. These boundaries form from the beginning of the graphene island growths and continue until the islands stop growing at full coverage. The second boundary type is the merged grain boundary. These boundaries are the same boundaries made when two islands meet. Their boundaries were not grown but form due to the lattices running out of space to continue. Using the interpreted map, measurements were made in these two categories and are shown in figure 2b. The results show that merged grain boundaries have a random distribution of lattice misorientations while grown boundaries have a preference of 14° with ten counts. This is expected since merged grain boundaries are formed by random lattice orientations meeting while grown grain boundaries are expected to have some sort of basis during its formation. This result confirms that the interpreted islands were correct.

Conclusion

This simple method that we have reported for analyzing the orientation of large CVD graphene grains provides a tool that can scale with the increasing quality of graphene. With it we have found important information about the graphene that we have grown such as grain size, grain shape, lattice orientation coverage, island positions, and grown and merged grain boundaries.

Acknowledgments

This work was supported by SWAN (GRC-NRI), AOARD-AFOSR (FA2385-10-1-4066) and the State of Texas ETF FUSION.

References

1. X. Li, W. Cai, J. An, S. Kim, J. Nah, D. Yang, R. Piner, A. Velamakanni, I. Jung, E. Tutuc, S. Banerjee, L. Colombo and R. Ruoff, *Science*, **324**, 1312 (2009).
2. P. Huang, C. Ruiz-Vargas, A. van der Zande, W. Whitney, M. Levendorf, J. Kevek, S. Garg, J. Alden, C. Hustedt, Y. Zhu, J. Park, P. McEuen and D. Muller, *Nature*, **469**, 389-392 (2011).

ECS Transactions, 45 (4) 83-88 (2012)
10.1149/1.3700456 ©The Electrochemical Society

Reduced perssure-chemical vapor deposition of high quality Ge layers on SiGe/Si superlayers for microelectronics and optoelectronics purposes

Da. Chen[a,b], Zhongying. Xue[a], Su. Liu[b], Miao. Zhang[a,*]

[a] State Key Laboratory of Functional Materials for Informatics, Shanghai Institute of Microsystem and Information Technology, Chinese Academy of Sciences, Shanghai, 200050, China
[b] School of Physical Science and Technology,Lanzhou University,Lanzhou 730000,China

We have investigated the structural properties of Ge thick films grown directly onto SiGe/Si superlayers/Si substrates using a production-compatible reduced pressure-chemical vapor deposition system. The thick Ge layers grown using germane and a low-temperature/high-temperature approach are in a tensile-strain configuration. X-ray-diffraction measurements showed that the Ge layer possesses a litter higher tensile strain as large as 0.31% than L/H conventional approach (0.17%), which is generated during the cooling from the high growth temperature due to the thermal-expansion mismatch between Ge and Si. In addition, we also exhibited the variations of Ge layers with changes of the superlayers. The process described in this letter for making high-quality Ge is uncomplicated and can be easily integrated with standard Si processes.

Introduction

The possibility of heteroepitaxial growth of pure germanium on silicon substrates has generated wide interest for the development of integrated electronic and optoelectronic devices (1-5). Due to its superior electron and hole mobility compared to Si, Ge has emerged as a feasible candidate to maintain performance requirements for future electronic applications (1). Furthermore, successful growth of Ge on Si provides a virtual substrate for the integration and fabrication of GaAs-based optical devices (2) and realization of III-V compound semiconductor metal-oxide-semiconductor-field-effect-transistor (MOSFET) on Si (3).

Because of the 4.2% lattice mismatch between Ge and Si, however, Ge epitaxial layers on Si (100) have high threading dislocation density (TDD) and surface roughness. Various methods were proposed to realize low TDD and smooth surface. In many cases, completely relaxed, thick graded Si_xGe_{1-x} layer was used for matching the lattice constant of the underlying layer to that of Ge, which would work to the disadvantage from manufacturing point of view, and cause a high threading dislocation density in the Ge layers and high surface roughness (6-7).

In this work, we demonstrate a method of combining SiGe/Si superlayers with deposition–annealing consisting of {LT Ge deposition/HT Ge deposition/HT H_2 annealing}. The obtained films were fully strain relaxed in whole 6in wafers and had smooth surfaces. We also exhibit the variations of Ge layers with changes of the superlayers.

83

Experiments

Samples with SiGe/Si superlayers were grown on Si (001) standard wafer by reduced pressure chemical vapor deposition (RPCVD). Pure silane (SiH_4) was used as the source of Si and germane (GeH_4) diluted at 10% in H_2 as the source of Ge, and the growth pressure was always 20 Torr. The substrates were 6 in. P-type Si wafers with resistivity in the range of 10-20 $\Omega \cdot cm$. The wafers were cleaned by Radio Corporation of American (RCA) method and loading into the growth chamber with robotic arm. At first, the wafers were baked at 1100°C in H_2 of 5 min to de-oxide, followed by Si buffer grown at 650°C to obtain a clean epi-ready surface. Secondly, SiGe/Si superlayers were grown at temperature 650°C. After that, 80-nm thick LT-Ge layer was deposited at 400°C. This LT-Ge layer was used to relaxes the strain in the Ge film without causing three-dimensional island nucleation which causes the surface roughness (8). Finally, nearly 780 nm HT-Ge layer was grown at 600°C and immediately after the epitaxial growth, hydrogen annealing was performed for 15 min 800°C. Fig. 1 illustrates the sample structures used in this study.

Strain relaxation in the Ge layers was analyzed by high resolution XRD and Raman spectroscopy equipped with an excitation wavelength of 514 nm. The crystalline quality of the Ge epilayer was characterized by TEM.

Results and discussion

Fig. 2 shows XRD spectra near Si (0 0 4) and Ge (0 0 4) peaks of L/H conventional Ge and L/H Ge with SLs. In the samples, the mean SiGe and Si layer thickness of the SLs were 4nm, 3nm, respectively. The apparent Ge concentration has fixed at 20%. The shift of peak of the Ge towards Si substrate indicates the degree of strain relaxation. It is found that tensile-strain configuration (R=105%) has been confirmed by computational simulation. L/H Ge with SLs and subsequent annealing has the equal relaxation degree compared to L/H sample with the identical thermal treatment, which is in agreement with Hartmann (9,13) and MIT researchers (10). The peak shift is caused by the difference of the thermal expansion coefficient of Si and Ge. During the cooling process from growth/annealing temperature down to room temperature, the originally fully relaxed Ge lattice that is tied to the Si substrate has to follow the slower contraction of the Si substrate, which causes a tensile strain in the Ge layer. In addition to the shift of Ge peaks, the full width at half maximum (FWHM) of the Ge peaks were 164s, 180s, respectively. The sample with SiGe/Si superlayers shows a little shoulder of increased intensity at the high-angle side of the Ge peak which is caused by Si diffusion into the Ge layer (11).

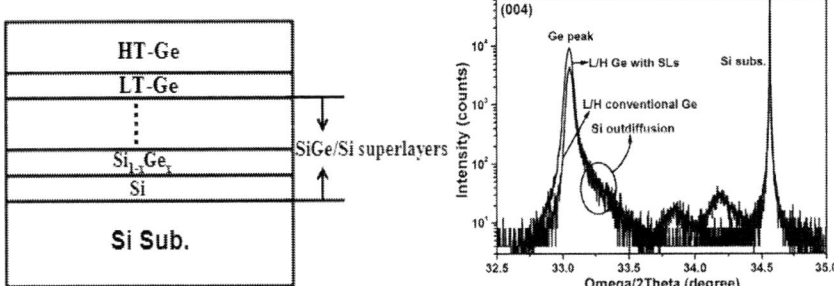

Fig.1 The SL-LH-Ge structure.

Fig.2 Omega-2Theta scans around the (004) diffraction order for the thick Ge layers grown using a 400°C /600°C process onto Si(001) substrates, one without any SLs interlayers, the other with a 5 times $Si_{0.8}Ge_{0.2}$/Si interlayers.

In order to discuss the the variations of Ge layers with changes of the superlayer in more detail, different circles and Ge concentration of SLs are discussed in use of XRD and Raman. Results are shown in Fig. 3 and Fig. 4. For each curves, three peaks originating from Si substrate, SiGe/Si superlayers, and Ge epilayer are clearly visible in fig. 3. Note that the full width at half maximum (FWHM) of the Ge peak decreases with increased circles and Ge concentration of SLs, implying that lower TDD were obtained in this way (TTD=$2\times10^{5}cm^{-2}$). As shown in Fig. 4, solid lines are experimental data and dotted straight lines are Ge-Ge mode from bulk Ge (ω=300.2cm^{-1}). The peak position is 299.5 cm^{-1} for samples. The value of in-plane strain can be estimated by using empirical expression (12):

$$\omega\left(cm^{-1}\right)= \omega_{Ge} - 400\varepsilon_{//} \qquad [1]$$

where ω and ω_{Ge} are, respectively, Raman shifts of Ge-Ge mode in epitaxial Ge layers and bulk Ge films, $\varepsilon_{//}$ the inplane strain in Ge epilayer. Through calculation, the stain for samples is almost 0.31%, which is a litter higher than that for L/H samples. The magnitude of strain is very similar to that evaluated from XRD data. With very little changes of Ge peak, it has been confirmed that SLs has some influences on epitaxial Ge layers.

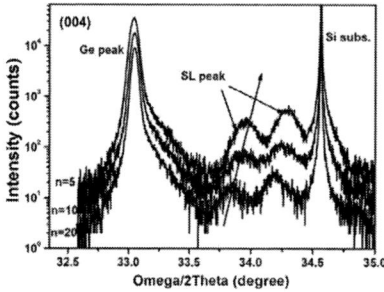 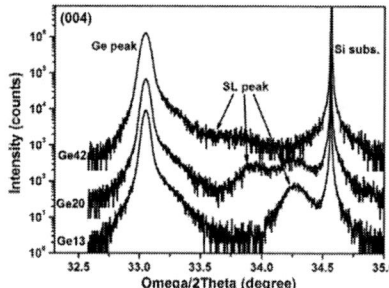

Fig.3 XRD images of SiGe SLs with different circles: n=5, 10, 20 and Ge concentration: Ge=13%, 20%, 42%.

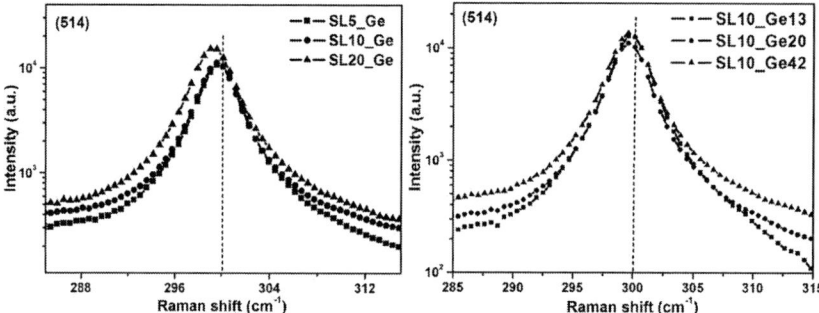

Fig.4 Raman images of SiGe SLs with different circles: n=5, 10, 20 and Ge concentration: Ge=13%, 20%, 42%.

We compared the location of dislocations with/without SLs in Ge layers using cross-section TEM images shown in Fig. 6. Fig. 6(a-b) were L/H-anneal Ge deposition sample, and Fig. 6(c-d) were L/H-anneal Ge sample with SLs (Ge concentration=42%). In case of Ge film without SLs, though most of the misfit dislocations (TD) are confined to the Ge/Si interface, some of them however propagate sideways into the LT Ge layer and present within 75nm of Ge. The TDD measured by mordant is in the range of 8-$10 \times 10^6 \text{cm}^{-2}$ and this value was measured uniformly across the entire surface. In contrast, by inserting SiGe/Si superlayers prior to the deposition of L-Ge, TDs are reduced but short dislocations are still visible in the superlayers (fig. 6c). The TDD of the Ge layer is so low that only few of them are visible on the Ge surface (TDD=$2 \times 10^5 \text{cm}^{-2}$). The contribution of low TDD in Ge layer can be divided into three parts: (1) the impact of SiGe/Si superlayer, that reduce the large lattice mismatch (4.2% at 300K). Furthermore, the presence of the numerous interfaces caused by SLs can influence the evolution of dislocations and change the bearing of trend. (2) high growth rate of Ge film in high temperature, that suppresses the propagation of the dislocations in the LT Ge layer. (3) high temperature annealing, that helps not only in smoothing the surface but also in causing the threading dislocations to propagate towards the edge of the substrate. This propagation is caused by linear thermal expansion coefficient difference ($\alpha_{Ge}= 6.0 \mu\text{m m}^{-1}\text{K}^{-1}$, $\alpha_{Si} = 2.6 \mu\text{m m}^{-1}\text{K}^{-1}$) (13).

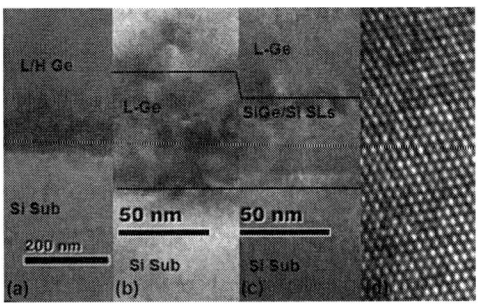

Fig.6 (a-b) Cross-section TEM images of L/H-anneal Ge deposition sample; (c-d) Cross-section TEM images of L/H-anneal Ge with SLs sample(Ge concentration = 42%).

Conclusion

We have grown high-quality, pure Ge layers on Si (001) substrates thanks to a method of combining SiGe/Si superlayers with deposition–annealing consisting of {LT Ge deposition/HT Ge deposition/HT H_2 annealing}. This method also reduces the total growth time. The obtained films were fully strain relaxed in whole 6in. wafers and had smooth surfaces. It not only promises the integration of bulk Ge devices on Si substrates, but also provides a platform for microelectronics and optoelectronics purposes.

Acknowledgments

This work was financially supported by the National Natural Science Foundation of China (Grant No. 61006088), the National Basic Research Program of China (973 Program) (Grant No. 2010CB832906), and the Natural Science Foundation of Shanghai (Grant No. 10ZR1436100).

References

1. C.O. Chui, S. Ramanathan, B.B. Triplett, P.C. McIntyre, K.C. Saraswat, *IEEE Electron Device Lett.*, **23**, 23 (2002).
2. J. Z. Li, J. M. Hydrick, J. S. Park, *J. Electrochem. Soc.*, **156**, H574 (2009).
3. M. M. Oye, D. Shahrjerdi, I. Ok, J. B. Hurst, S. D. Lewis, S. Dey, D. Q. Kelly, S. Joshi, X. Yu, M. A. Wistey, J. J. S. Harris, J. A. L. Holmes, *J. Vac. Sci. Technol. B.*, **25**, 1098 (2007).
4. H. Zhai, Z. Zhang, L. Li, S. Ma, *Opt Quant Electron*, **41**, 957 (2009).
5. K. W. Ang, M. B. Yu, G. Q. Lo and D. L. Kwong, *IEEE Electron Device Lett.*, **29**, 1124 (2008).
6. H. C. Luan, D. R. Lim, K. K. Lee, K. M. Chen, J. G. Sandland, K. Wada, and L. C. Kimerling, *Appl. Phys. Lett.*, **75**, 2909 (1999).
7. G. L. Luo, T. H. Yang, E. Y. Chang, C. Y. Chang and K. A. Chao, *J. Appl. Phys.*, **42**, L517 (2003).

8. L. Colace, G. Masini, G. Assanto, G. Capellini, L. Di Gaspare, E. Palange, F. Evangelisti, *Appl. Phys. Lett.*, **72**, 3175 (1998).
9. J. M. Hartmann, J. F. Damlencourt, Y. Bogumilowicz, P. Holliger, G. Rolland, T. Billon, *J. Crystal Growth*, **90**, 274 (2005).
10. Y. Ishikawa, K. Wada, D. D. Cannon, J. Liu, H. C. Luan, L. C. Kimerling, *Appl. Phys. Lett.*, **82**, 2044 (2003).
11. Y. J. Yamamoto, P. Zzumseil, T. Arguirov, M. Kittler, B. Tillack, *Solid-State Electronics*, **60**, 2 (2010).
12. P. H. Tan, K. Brunner, D. Bougeard, G. Abstreiter, *Phys. Rev. B*, **68**, 5302 (2003).
13. J. M. Hartmann, A. Abbadie, A. M. Papon, P. Holliger, G. Rolland, T. Billon, J. M. Fedeli, M. Rouviere, L. Vivien, S. Laval, *J. Appl. Phys.*, **95**, 5905 (2004).

CHAPTER 4

GE AND III/V TECHNOLOGIES

III-Sb MOSFETs: Opportunities and Challenges

Aneesh Nainani[1,2], Ze Yuan[1], Archana Kumar[1], Brian R. Bennett[3], J. Brad Boos[3], Krishna C. Saraswat[1]

[1]Center of Integrated Systems, Stanford University, Stanford, California 94305, USA
[2]Applied Materials, Sunnyvale, California 94085, USA
[3]Naval Research Laboratory, Washington, DC 20375, USA
Email: nainani@stanford.edu

This papers reports on our work on III-Sb based MOSFETs. We demonstrate that heterostructure MOSFETs using InGaSb as the channel material are capable of enabling a high performance complementary MOS technology in a single channel material.

Introduction

Antimony (Sb) based compound semiconductor materials have the highest electron and hole mobilities amongst all compound semiconductor materials. Transistors using Sb-channel deliver much higher performance at significantly lower power. As compared to $In_xGa_{1-x}As/InP$ system Sb's have higher band offsets for a heterostructure design and higher mobility for holes. Sb-based channel materials are the only candidates, which can offer a complementary technology outperforming silicon in a single channel material. A CMOS technology in a single channel material is more advantageous for integration vs. one which requires two different materials with different lattice constants for the nMOS and pMOS. Besides this the ~6.1Å InAs-GaSb-AlSb lattice system offsets rich possibilities for band engineering as shown in Figure. 1, which can be applied to other applications such as tunnel FETs as well.

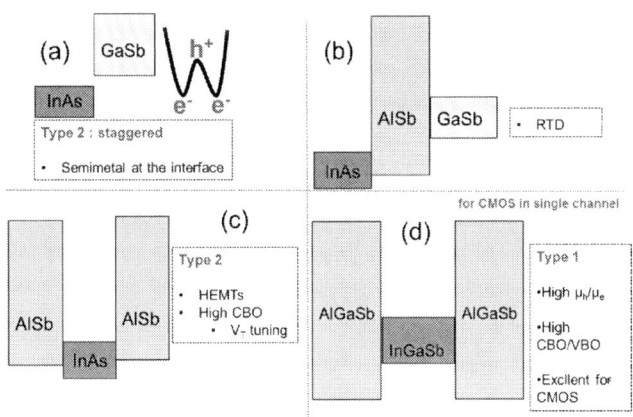

Figure 1: The 6.1-6.2Å lattice system of InAs-AlSb-GaSb offers large latitude for band engineering. Cartoon above shows the various band lineups possible. InGaSb-AlGaSb system has the potential to enable CMOS technology in a single channel material (d).

Opportunities & Challenges

Schottky-gate FET devices with $In_xGa_{1-x}Sb$ channel have achieved f_T of 305GHz at 0.5V V_{DS} (L_G=85nm) for n-channel (1) & f_T of 140GHz (L_G=40nm) for p-channel (2). While excellent f_T and f_{MAX} values have been reported in HEMT like Sb-channel devices the ON current of these devices has been limited due to leakage current through the Schottky gate and large access resistance. Also while Sb-channel devices with gate lengths down to 40nm have been demonstrated, the source/drain (S/D) separation has remained in micrometer scale due to the use of alloyed contacts for S/D. Figure. 2, contrasts our approach against the HEMT work. In our work we have focused on a self-aligned process with high-k dielectric between the heterostructure and metal gate to cut down the gate leakage while preserving the high mobility in the channel. Low resistance source/drain contacts right next to the gate reduces in the access resistance and enables making devices with short-pitch suitable for large-scale integration. In this paper we report on our work on the quantum well MOSFETs using Sb-based III-V semiconductors with $In_xGa_{1-x}Sb$ as the channel material which can be used for making both high performance n- and p- channel transistors.

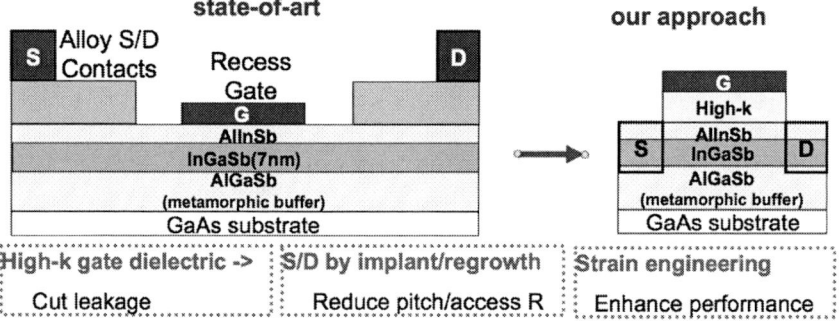

Figure 2: Sb-channel MOSFETs featuring a self-aligned source/drain with high-dielectric and using strain engineering to enhance performance

Figure 3 plots the band lineup of the $In_xGa_{1-x}Sb/Al_yGa_{1-y}Sb$ system. The bandgap and the conduction / valence band offsets in the system were measured using PL and XPS measurements (3). Also this was coupled with a modeling exercise with bandstructure calculated using tight binding method which proved that heterostructure design has sufficient conduction and valence band offsets for confining the electrons and holes in the high mobility $In_xGa_{1-x}Sb$ channel while preventing them from spilling over into the $Al_yGa_{1-y}Sb$ barrier layer which has low mobility (Figure. 3).

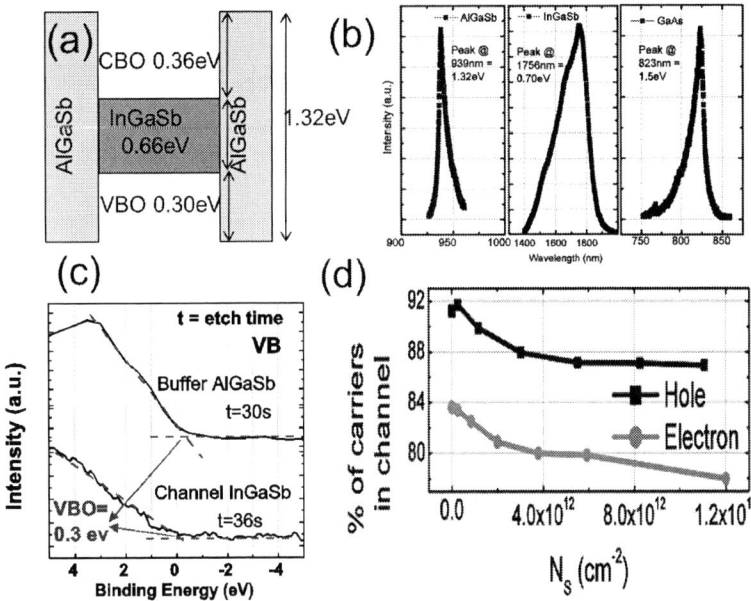

Figure 3: For the heterostructure design with InGaSb channel and AlGaSb barrier (a), the bandgap and conduction/valence band offsets (CBO/VBO) were measured experimentally using (b) PL and (c) XPS measurements. Results from bandstructure calculations using tight binding method (d) predict that this heterostructure design can confine majority of the carrier in the channel while preventing them from spilling over into the barrier layer even at high sheet charge.

The main challenge in the realization of Sb-MOSFETs has been the highly reactive nature of the Sb-surface and lack of known solutions for source/drain technology for Sb-based channels (Figure 2). Recently utilizing sophisticated surface analysis techniques such as low energy radiation from synchrotron we developed an optimal chemical clean to remove the native oxides and render a stable Sb-surface (4). Subsequent deposition of ALD dielectric resulted in excellent CV characteristics with mid-bandgap D_{it} of $3 \times 10^{11}/cm^2eV$ as shown in Figure 4. This high quality dielectric coupled with S/D formation by ion implantation and strain engineering resulted in Sb-channel pMOSFETs with subthreshold swing (SS) of 120mV/decade and $I_{ON}/I_{OFF} > 10^4$ as shown in Figure 5 (5). Use of high-k gate electric ensured that the gate leakage current remained orders of magnitudes lower as compared to source/drain currents (Figure 5). We also demonstrated that the complete processing of these MOSFETs can be done at temperature below 350°C, which is advantageous for heterogeneous integration on a Si CMOS platform (6).

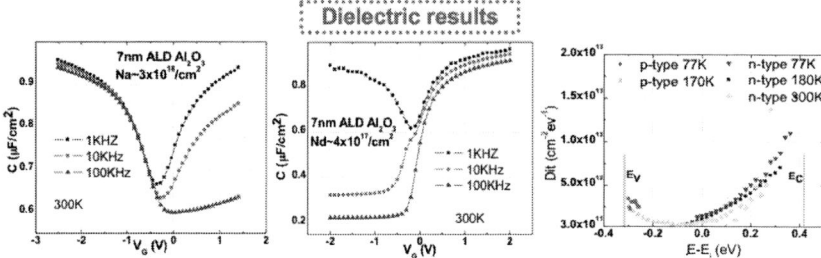

Figure 4: Measured CV's from 1kHz-1MHz on p-type (left) and n-type (center) GaSb at 300K. D_{it} distribution (right) was measured across the bandgap using conductance method. Mid bandgap D_{it} of $3 \times 10^{11} \text{cm}^{-2}\text{eV}^{-1}$ was achieved

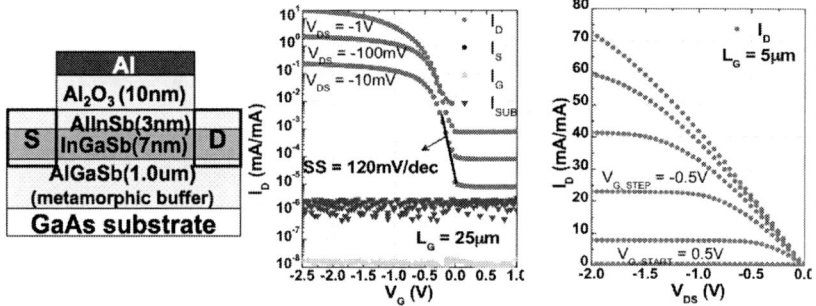

Figure 5: I_D-V_G and I_D-V_D in our buried channel pMOSFET device. SS is 120mV/decade while I_G and I_{SUB} remains orders of magnitude lower compared to I_D/I_S

Hole mobility in our buried (surface) channel devices was > 100 (50) % higher than germanium MOSFETs even at high sheet charge (5). Although our process and heterostructure design is suited for both p/n-channel devices, pMOSFET was attempted first as it has been an impediment in implementation of III-V complementary logic. We have recently achieved electron mobility exceeding $6000\text{cm}^2/\text{Vs}$ in the same channel material used for pMOSFET and are working towards demonstrating a complementary technology in single channel material outperforming silicon (Figure 6) (3).

Some of the challenges we foresee in implementing a nMOSFET in the same system are poor activation of n-type dopants in $In_xGa_{1-x}Sb$ & high contact resistivity for nMOSFET due to Fermi level pinning near the valence band edge in these materials (Figure 7(a)). The electron Schottky barrier height can be reduced by inserting ultrathin dielectrics at the metal-semiconductor interface to unpin the Fermi level. However, this technique introduces tunneling resistance from the large conduction band offset (CBO) between the insulator and $In_xGa_{1-x}Sb$ (Figure 7(b)). We have figured out a way to substantially reduce the contact resistance to n-type $GaSb/In_xGa_{1-x}Sb$ by introducing a TiO_2 interfacial layer between the semiconductor which unpins the Fermi level while introducing little tunneling resistance due to near zero CBO of TiO_2 with GaSb (Figure

7(c)). In our preliminary results nearly 4 orders of magnitude increase in current density was observed for contacts to n-GaSb using interfacial TiO_2 (Figure 7(d)) (7).

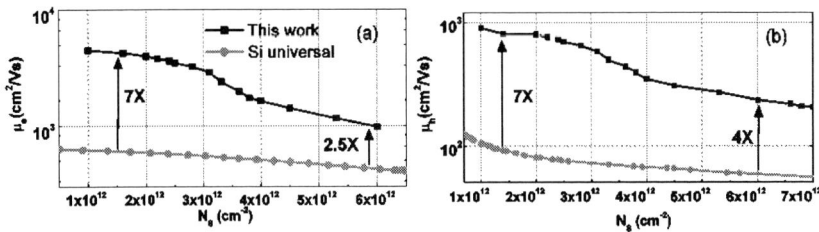

Figure 6: (a) Electron and (b) hole mobility measured in $In_xGa_{1-x}Sb$ channel vs. N_s using gated Hall measurement. Electron/hole mobility is > 2.5X/4X as compared to Si universal even at high N_s.

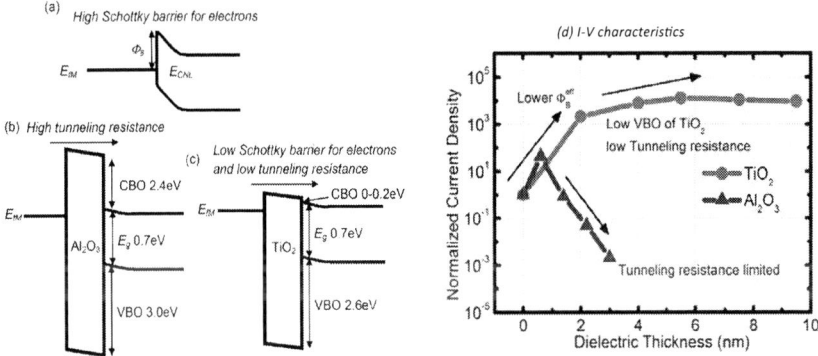

Figure 7: (a) Pinning increases Schottky barrier to electrons. (b) Insertion of Al_2O_3 or (c) TiO_2 unpins the Fermi level near the GaSb valence band while adding minimal tunneling resistance in the case of TiO_2. (d) Nearly 4 orders of magnitude increase in current density is observed for TiO_2 contacts to n-GaSb. (7)

This technique can be utilized to fabricate MOSFEs with Schottky contact S/D without the need of a diffused junction or alloyed contacts. Figure 8, shows such one such scheme, for PMOS the pinned metal Fermi level at the valance band and for NMOS the unpinned Fermi level due to insertion of TiO_2 will give low contact resistance.

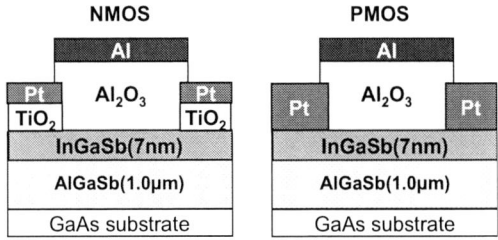

Figure 8: Complementary CMOS with Schottky contact S/D.

Conclusions and summary

In summary, we have proposed InGaSb as the channel material which can enable III-V CMOS in a single channel material. Electron/hole mobility > 4000/900cm^2/Vs was achieved in the same heterostructure stack. Excellent pMOSFET characteristics were demonstrated with I_{ON}/I_{OFF} of > 10^4 and SS of 120mV/decade. Contact resistance is identified as the main challenge for InGaSb NMOS and we show that this can be much improved with a TiO$_2$ interfacial layer. On a concluding remark we must mention that heterostructures using Sb-based semiconductors are also leading candidates for enabling tunnel FETs which achieve high ON current and are being explored by several research groups (8-9).

Acknowledgments

We would like to acknowledge the Stanford INMP program, Applied Materials, SEMATECH and Intel Corporation and Office of Naval Research for funding parts of this research at Stanford University and Naval Research Laboratory. Aneesh Nainani is additionally thankful to Intel Corporation for a PhD fellowship.

References

1. S. Datta, T. Ashley, J. Brask, L. Buckle, M. Doczy, M. Emeny, D. Hayes, K. Hilton, R. Jefferies, T. Martin, T. J. Phillips, D. Wallis, P. Wilding, and R. Chau, in *Technical Digest - International Electron Devices Meeting*, p. 763 (2005).
2. M. Radosavljevic, T. Ashley, A. Andreev, S. D. Coomber, G. Dewey, M. T. Emeny, M. Fearn, D. G. Hayes, K. P. Hilton, M. K. Hudait, R. Jefferies, T. Martin, R. Pillarisetty, W. Rachmady, T. Rakshit, S. J. Smith, M. J. Uren, D. J. Wallis, P. J. Wilding, and R. Chau, in *Technical Digest - International Electron Devices Meeting.* p. 727 (2008).
3. Z. Yuan, A. Nainani, B. R. Bennett, J. B. Boos, M. G. Ancona, and K. C. Saraswat in *2011 International Semiconductor Device Research Symposium (ISDRS).*
4. A. Nainani, Y. Sun, T. Irisawa, Z. Yuan, M. Kobayashi, P. Pianetta, B. R. Bennett, J. B. Boos, and K. C. Saraswat, in *Journal of Applied Physics* **109**, 114908 (2011)
5. A. Nainani, Z. Yuan, T. Krishnamohan, B. R. Bennett, J. B. Boos, M. Reason, Ma. G. Ancona, Y. Nishi, and K. C. Saraswat, in *Journal of Applied Physics* **110**, 014503 (2011)
6. A. Nainani, T. Irisawa, Z. Yuan, B.R. Bennett, J.B. Boos, Y. Nishi, and K.C Saraswat, in *Transactions on Electron Devices*, **58**, 10, pp. 3407-3415, Oct. 2011
7. Z. Yuan, A. Nainani, Y. Sun, J.-Y. Lin, P. Pianetta, and K. C. Saraswat, in *Applied. Physics Letters* **98**, 172106 (2011)
8. J. Knoch, and J. Appenzeller, in *Electron Device Letters*, **31**, 4, pp.305, April 2010
9. D.K. Mohata, R. Bijesh, S. Mujumdar, C. Eaton, R. Engel-Herbert, T. Mayer, V. Narayanan, J.M. Fastenau, D. Loubychev, A. K. Liu, and S. Datta, in *Technical Digest - International Electron Devices Meeting*, pp. 33.5.1-33.5.4 (2011)

Passivation challenges with Ge and III/V devices

S. Sioncke[a], D. Lin[a], L. Nyns[a], A. Delabie[a], A. Thean[a], N. Horiguchi[a], H. Struyf[a], S. De Gendt[a,b], M. Caymax[a]

[a]Imec, Kapeldreef 75, B-3001 Leuven, Belgium
[b]Department of Chemistry, K.U. Leuven, Celestijnenlaan 200F, B-3001 Leuven, Belgium

> Ge and III/V materials are investigated as high mobility channel materials for devices beyond the 14 nm technology node. Passivation of the interface is a prerequisite as dangling bond states trap charges and reduce the mobility. Passivation of the Ge/high-κ interface can be done by using GeO_2 as a passivation layer. However, large thicknesses are needed to keep a low defect density. Introduction of S at the interface can solve this issue. Passivation of III/V materials is still a major problem. Oxidation of the III-As surface leads to stress at the surface creating As-As bonds. Oxides can be partially removed by wet treatment or during the ALD deposition. The defects related to As-As bonds cannot be removed by the ALD process. We demonstrate that also for InP, an oxide free interface does not solve the passivation problem.

The passivation problem: introduction.

Replacement of the Si channel by high mobility materials such as Ge and III/V opens a lot of new research topics for materials scientists and device engineers. The high intrinsic hole mobilities of Ge and electron mobilities of III/V materials will only pay off if they can be demonstrated in a Ge or III/V channel device that is co-integrated on a Si platform. One of the most challenging tasks is to grow these materials on a large scale on Si bulk wafers. Another problem that needs to be tackled is the passivation of the interface between the high mobility channel and the high-κ. The bulk properties are interrupted at the interface and this introduces stress which in itself results in defect states trapping the carriers and lowering the mobility of the channel. Simply stated, passivation should preserve the bulk properties close to the interface.

The last decade, a lot of progress has been made in passivating the Ge/high-κ interface. Several routes have been explored: Si passivation, GeO_2, GeO_xN_y, S-passivation ... [1, 2, 3, 4, 5, 6, 7, 8]. GeO_2 passivation is the most promising candidate as the lowest D_{it} (Density of interface states) is achieved by this route. Recently, pMOS and nMOS devices with high mobilities have been demonstrated in combination with low EOT (Equivalent Oxide Thickness) [9, 10, 11].

On III/V materials, the passivation problem appears to remain a bottleneck. A more fundamental understanding has been built up throughout the last years. However, passivation of the III/V/high-κ interface seems to be even more challenging than passivation of the Ge/high-κ interface. GaAs shows several defect states in the band gap [12]. The states at the band edges are reduced to some extent by changing the pretreatment. They are related to Ga-O, As-O, Ga- or As-dangling bonds and can be reduced by the pretreatment or during the ALD (Atomic Layer Deposition). The mid gap defect states most likely originate from As and Ga vacancies caused by the physical stress of the amorphization of the interface. These vacancies originate from the oxidation process of the surface and cannot be removed by removal of the oxides [13]. On InP, it is unclear if the same defects might appear after oxidation of the surface. Besides the passivation of the interface between III/V and the high-κ oxide, passivation of border traps should also be considered on III/V materials. The low conduction band density of states is responsible for the fact that the surface Fermi level can be biased well into the conduction band [14, 15]. This raises the probability of charge trapping action at accumulation between the oxide border traps and conduction band electrons. Any defects in the oxide will therefore also play a role in the final device performance.

Ge passivation

GeO$_2$-passivation of the Ge gate stack.

Passivation of the Ge/high-κ interface has been investigated intensively over the last decade and a lot of progress has been reported. Several passivation routes have been explored and the most promising candidates are Si and S passivation for pMOS applications and GeO$_x$(S$_y$) for nMOS applications. Si passivation is achieved by the epitaxial growth of a few monolayers of Si on a clean Ge substrate. This layer is partially oxidized to SiO$_2$. This route showed promising pMOS mobilities [1]. However, due to a high defect density at the conduction band edge, this passivation cannot be used for nMOS applications. GeO$_2$ passivation solves this problem. Scientists believed that the beauty of the Si/SiO$_2$ interface could not be reproduced on Ge as GeO$_2$ suffers from a low thermal stability and is also highly water soluble. Therefore, the initial tendency was to remove the GeO$_2$ before gate stack deposition. However, this resulted in very bad passivation of the interface [16]. First-principle calculations on the (100)Ge/GeO$_2$ interface by Houssa et al. [17] showed that the initial stress at the semiconductor/oxide interface is caused by the stress due to a lattice mismatch between the semiconductor crystal and its amorphous oxide, the stress being proportional to the Young modulus of the oxide. The Young modulus of GeO$_2$ is about 1.7 times lower than the Young modulus of SiO$_2$, resulting in a lower amount of initial stress at the Ge/GeO$_2$ interface. One could therefore expect that at low oxidation temperatures the corresponding defect densities due to dangling bonds would be lower for the Ge/GeO$_2$ interface compared to the Si/SiO$_2$ interface. Stress relaxation can be obtained by viscous flow of the oxide. The viscosity as well as the activation energy for viscous flow is much lower for GeO$_2$ compared to SiO$_2$, allowing for stress relaxation at much reduced temperatures. Stress relaxation at the Ge/GeO$_2$ interface is expected between 400C-600C. While 800C-1000C is needed for similar stress reduction at the Si/SiO$_2$ interface. However, here we are approaching the major problem of the Ge/GeO$_2$ interface. At 400C, a reaction starts to take place at the interface between Ge and GeO$_2$:

$$Ge(s)+GeO_2(s) => GeO\ (g)$$

GeO is volatile at 400C and desorbs from the interface, resulting in a high defect density. Growing the GeO$_2$ at high temperatures (550C) and high O$_2$ pressures (>1atm), can suppress this decomposition reaction and results in very high quality gate stacks [2,18, 19] showing the potential of this gate stack for nMOS and pMOS applications. As the GeO$_2$ growth increases with increasing temperature and O$_2$ pressure, the resulting stacks are very thick. The high quality interfaces reported were always for very thick GeO_2 films (~20 nm). Therefore, this raises the question of how to scale these gate stacks. However, recent results were published using GeO_x as a passivation layer [9]. A GeO_x interfacial layer was achieved by electron cyclotron resonance (ECR) plasma post oxidation of the Al_2O_3/Ge structure. Zhang et al. demonstrated that a 0.5 nm GeO_x is sufficient to achieve low defect densities ($1\ 10^{11}$ - $2\ 10^{11}$ /cm^2eV at E_i-0.2eV). This GeO_x stack corresponds to a total EOT thickness of 1.1 nm. Thinner GeO_x resulted in an increase of defects. However, the reported D_{it} levels at this EOT ranged between $1\ 10^{11}$ and $9\ 10^{11}$ /cm^2eV.

S-passivation of the Ge gate stack.

The oxide free Ge gate stack. Recently, we have demonstrated high quality stacks when introducing S at the Ge/high-κ interface. The initial idea was to avoid any O at the interface. By introducing S, a higher flexibility from the Ge-S bond compared to the Ge-O bond could reduce the stress at the Ge/high-κ interface. In [20] a Young modulus of 11.76 GPa for GeS_2 was reported. This young modulus is about 4x lower than the Young modulus of GeO_2. According to Houssa et al. the initial stress at the interface should scale linearly to the Young modulus of the oxide. If this would also apply to the Ge/GeS_2 interface, this means an initial stress reduction of a factor ~4. Although the viscosity is higher than for GeO_2 ($\eta GeO_2=6\ 10^{-8}$ P, $\eta GeS_2=1\ 10^{(-3.73)}$P), the activation energy for viscous flow is lower ($E_aGeO_2=3.3$ eV, $E_aGeS_2=2.7$ eV) [21]. Stress reduction should therefore also occur at even lower temperatures than for GeO_2.

A S layer can be achieved by wet treatment by using $(NH_4)_2S$ [22,7] or by gas phase exposure of the clean Ge substrate to H_2S [8]. Al_2O_3 was deposited by ALD using Tri-MethylAluminum (TMA) and H_2O. For both $(NH_4)_2S$ and H_2S, high quality gate stacks for pMOS applications were obtained: low defect densities at the valence band edge (~1 $10^{11}cm^2$/Vs) but increasing towards the conduction band edge. The CV (Capacitance-Voltage) data for the Ge/S/Al_2O_3 gate stacks on p- and n-type Ge are shown in Figure 1.

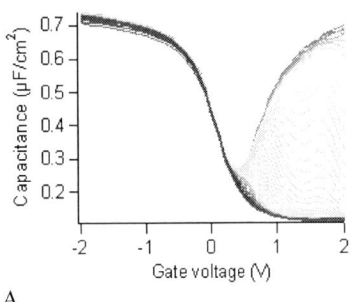

A

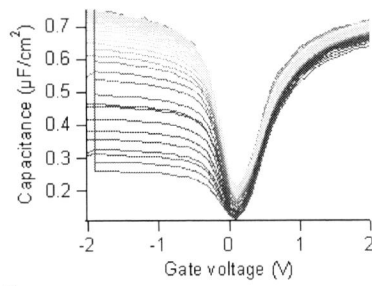

B

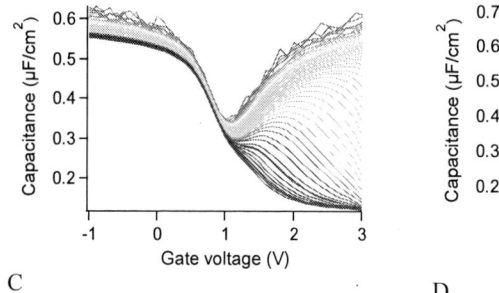 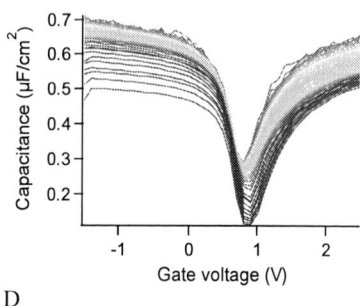

C   D

Figure 1. C-V measurements for the $Ge/S/Al_2O_3$ (TMA/H_2O) gate stack with Pt dots after FGA anneal for A-C) p-type Ge and B-D) n-type Ge substrates. 30 frequencies were varied logarithmically between 100 Hz (upper line) and 1 MHz (lower line). The measurements were carried out at room temperature. S-passivation was achieved by using $(NH_4)_2S$ (A-B) or H_2S gas exposure (C-D).

Moreover, XPS analysis of the gate stacks showed that the interface was oxide free (see Figure 2). NEXAFS (Near Edge X-ray Absorption Fine Structure) spectroscopy showed that Ge-S-Al bonds were present at the interface [7,8].

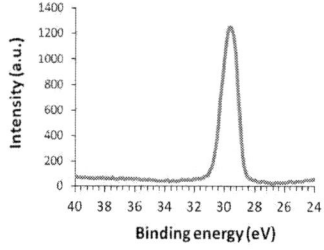 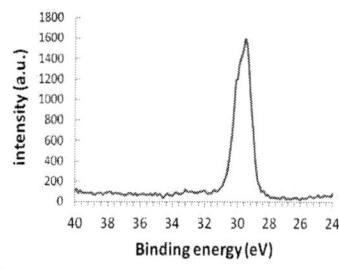

A   B

Figure 2. XPS spectra of the Ge3d peak at 28 degrees from the normal after several cycles of TMA/H_2O on a Ge/S substrate. The Ge/S substrate was obtained by dipping the substrate in $(NH_4)_2S$ (A) or by exposing the clean Ge substrate to H_2S (B).

As no GeO_2 is present, this gate stack exhibits high scaling potential. However, in order to introduce this gate stack into the 14 nm technology node, the EOT should be lowered to sub 1 nm EOT. Changing the high-κ oxide to HfO_2 or ZrO_2 did not result in a gate stack with similar defect densities. The same trend was still observed: low defect densities at the valence band edge and increasing defect densities towards the conduction band edge. However, the values are shifted up to 1 order of magnitude [7]. Therefore, another approach needs to be followed: bi-layers were made consisting of $Ge/S/Al_2O_3/HfO_2$. The scaling limit is determined by the layer closure of the Al_2O_3. Growth behavior of the Al_2O_3 (TMA/H_2O) was extensively investigated [8, 23, 24]. XPS revealed a slow layer closure (~20 cycles). It was shown that layers are not closed up to 10 cycles of TMA/H_2O: GeO_2 is present in the Ge3d spectrum due to re-oxidation through the holes in the layer. From 20 cycles on, no Ge-oxides are detected. Therefore, we concluded that the point of layer closure is situated around 20 cycles. This means that the scaling limit is determined by the minimum amount of TMA/H_2O cycles needed to close the Al_2O_3 layer. 20 cycles of TMA/H_2O results in a ~1.5 nm thick Al_2O_3. The bi-layer approach was demonstrated using 20 cycles of Al_2O_3 on this S-treated Ge surface followed by a 2 nm HfO_2 deposition. This resulted in a device with high pMOSFET mobility (>200 cm^2/Vs) at an EOT of 1.5 nm (Figure 3).

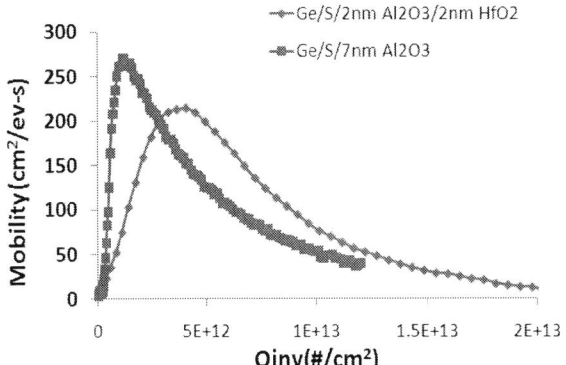

Figure 3. Mobility (cm^2/Vs) as a function of inversion charges (#/cm^2) for the Ge/S/Al_2O_3 (7 nm) and the Ge/S/Al_2O_3 (2nm)/HfO_2 (2 nm) stack. The S-terminated Ge surface was obtained by a dip in $(NH_4)_2S$.

The GeO_xS_y passivation layer. Depositing Al_2O_3 by TMA/H_2O resulted in an oxide free interface with low defect densities at the valence band edge. However, switching the oxidant precursor to O_3 results in an interface very similar to GeO_2 [8]. Low defect densities are observed at the conduction band edge (Figure 4). The resulting oxide is very thin (~0.8 nm) as determined by XPS shown in Figure 5. A comparable stack was obtained by deposition of TMA/O_3 on an HF cleaned Ge surface. The resulting oxide thickness was very comparable to the stack with a S pretreatment (~0.8 nm). The defect density as a function of the position in the band gap is also shown in Figure 4. This is a direct comparison between the GeO_xS_y and GeO_x passivation. It is clear that the introduction of S into the GeO_x results in a defect density at the interface that is lowered by a factor of ~5. S is still present at the interface as confirmed by TOFSIMS (Time of Flight Secondary Ion Mass Spectroscopy). By using NEXAFS, we could resolve the S-bonding states and after Al_2O_3 deposition, a clear oxidized S state (+6) is present besides the Ge-S-Al bonds. Therefore, we concluded that S is mixed into the thin Ge-oxide forming a GeO_xS_y passivation layer. Thinning the GeO_2 below 2 nm usually results in a degraded interface [25]. With this thin (<1nm) GeO_xS_y, we are able to achieve a high quality gate for nMOS applications with aggressive scaling possibilities.

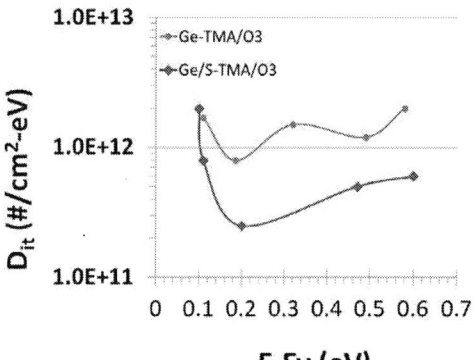

E-Ev (eV)

Figure 4. D_{it} distribution in the Ge band gap for the Ge/S/Al$_2$O$_3$ (10 nm) and Ge/Al$_2$O$_3$ (10 nm) where Al$_2$O$_3$ is deposited by TMA/O$_3$. The S-terminated Ge surface was obtained by exposing the clean Ge surface to H$_2$S. The D_{it} is extracted from the parallel equivalent conductance G_p [26] obtained at RT, 77 K and 160 K. Pt dots are used as gate metal and a FGA was performed at 400 C for 10 min.

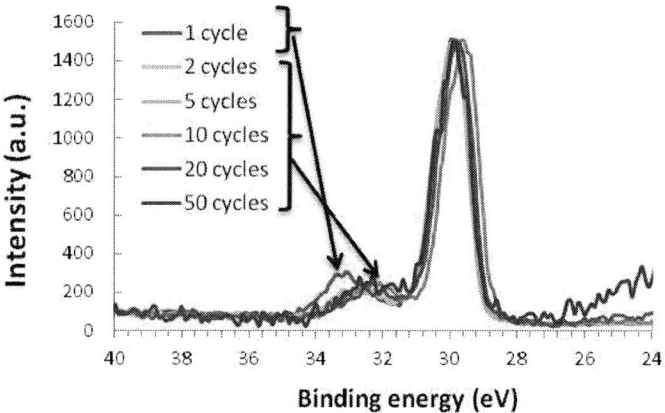

Binding energy (eV)

Figure 5. XPS data in the Ge3d region of the Ge/S/TMA-O$_3$ sample at an angle of 21.88 degrees. The S-terminated Ge surface was obtained by exposing the clean Ge surface to H$_2$S.

A bi-layer approach using a 2 nm Al$_2$O$_3$ (TMA/O$_3$) layer combined with a 2 nm HfO$_2$ layer was investigated. The resulting D_{it} contributions obtained from G-V data using the conductance technique [26] are shown in Figure 6. It is clear that the bi-layer approach did not degrade the interface. Further reduction of the EOT is possible by reducing both the Al$_2$O$_3$ and HfO$_2$ thickness. As the GeO$_x$S$_y$ is formed from the 2nd cycle as can be seen from XPS data shown in Figure 5, it could be possible that reduction of the Al$_2$O$_3$ is possible down to 2 cycles of TMA/O₃. In Figure 7, the CV is shown for the gate stack Ge/GeₓSᵧ/Al₂O₃ (TMA/O₃=2 cycles)/HfO₂ (HfCl₄/O₃=16 cycles)/TiW. The gate stack received a FGA anneal at 400 C for 10 min. The C_{acc} measured is 3.8 $\mu F/cm^2$ and corresponds to an EOT of 0.6 nm. This highly scaled gate stack shows clear traces of gate leakage in the CV. Therefore, it is difficult to extract the quality of the resulting interface from the CV only. Therefore we are currently investigating the quality of this aggressively scaled gate stack by measuring the resulting nMOS devices.

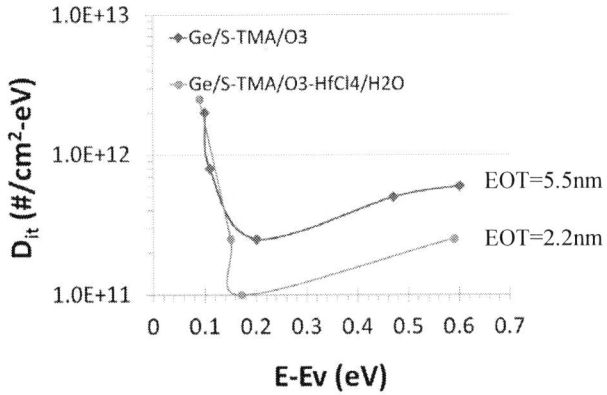

Figure 6: D_{it} distribution in the Ge band gap for the Ge/S/Al₂O₃ (10 nm) and Ge/S/Al₂O₃ (2nm)/HfO₂ (2nm) where Al₂O₃ is deposited by TMA/O₃, the HfO₂ layer is deposited by HfCl₄/H₂O. The D_{it} is extracted from the parallel equivalent conductance G_p [26] obtained at RT, 77 K and 160 K. Pt dots are used as gate metal and a FGA was performed at 400 C for 10 min.

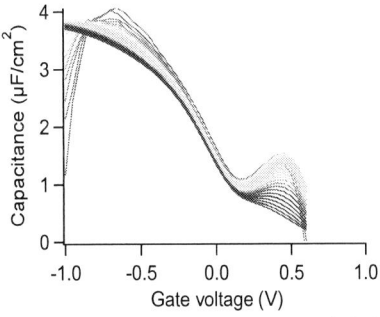

Figure 7: CV of the gate stack p-Ge/GeₓSᵧ/Al₂O₃(TMA/O₃=2cycles)/HfO₂ (HfCl₄/O₃=16cycles)/TiW after FGA anneal at 400 C. 30 frequencies were varied logarithmically between 100 Hz (upper line) and 1 MHz (lower line).

Passivation of the III/V gate stack: going beyond interface passivation.

Passivation of the $In_xGa_{1-x}As$/oxide interface.

It has been shown by Scarrozza et al. that oxidation of the GaAs surface should be avoided as this introduces stress at the surface with the creation of As-As bonds which act as traps [13]. Therefore, the removal of these oxides by any means (wet chemical or an ALD selfcleaning process) can remove the defects related to Ga-O or As-O bonds but this cannot remove the defects created during the oxidation process prior to the treatment. Experimental evidence was reported by several groups. Brammertz et al. [12] reported defect states close to the conduction band edge and valence band edge and related these to Ga-O, Ga dangling bonds and As-O, As dangling bonds respectively. At mid gap, donor and acceptor peaks were detected. An $(NH_4)_2S$ treatment prior to ALD or a forming gas anneal after ALD are able to reduce the defects at the band edges. However, the mid gap peaks are not affected by these treatments. The peaks at mid gap were related to Ga-Ga and As-As bonds: both structural defects in the surface caused by the oxidation of the surface. The same conclusions could be made for $In_{0.53}Ga_{0.47}As$. However, the peak distribution is shifted compared to GaAs: a large peak is observed close to the valence band edge and a smaller peak exists at mid gap. Both peaks are donor peaks and are uncharged when the Fermi level is close to the conduction band edge which allows for a high I_{on}. However, the presence of the peaks will affect the I_{off} substantially.

By using in-situ XPS on a S-treated GaAs surface Wang et al [27], reported the reduction of Ga-oxides and As-oxides by ALD of Al_2O_3 (TMA/H_2O). However, the As-As bonds persisted throughout the ALD cycles. Similar results were also found on $In_xGa_{1-x}As$ (0<x<1) (100) surfaces [28, 29]. Also Tallarida et al [30] reported the reduction of Ga- and As-oxides by ALD of Al_2O_3 (TMA/H_2O). Ga^{3+} states were completely removed. The Ga-suboxides ($Ga^{(1\pm\Delta)}$) remained unchanged. As^{3+} was completely reduced, while the As suboxides changed in intensity and binding energy, probably because of the formation of As-O-Al bonds. However, the position of the Fermi level did not change substantially during the TMA exposure. This indicates that interface defects were not significantly reduced during the TMA cleaning although the high valence As- and Ga-oxides were removed.

In [31] defect levels at the $In_xGa_{1-x}As$/oxide and InP/oxide interface were simulated through hybrid density functional calculations. They found that cation (In, Ga) dangling bonds are close to the conduction band edge (GaAs) or in resonance with the conduction band ($In_{0.53}Ga_{0.47}As$, InAs).The As dangling bond lies slightly above the valence band edge for all compositions, while the P dangling bond lies at mid gap. This finding suggests again that the dangling bonds are responsible for the states at the band edges which can be reduced during ALD self cleaning or by a pretreatment (NH_4OH, $(NH_4)_2S$, Si passivation) or a FGA. They also considered defects occurring in GaAs from As antisites. The defect levels calculated were in good agreement with the experimental defect levels around mid gap energies. The key for passivating these interfaces is therefore finding a way to avoid structural damage of the III/V surface before and during the high-κ deposition.

A defect free InP surface by gas phase HCl etching.

Although a lot of progress has been made in understanding the origin of defects at III-V/oxide interfaces, the origin of defects at the InP/oxide interface are not well-understood. Lately, some simulations have been done to investigate the possible defects at the InP/oxide interface and the corresponding levels in the band gap. Robertson et al. [32] showed that the P-P bond, analogous to the As-As bonds in InGaAs, lies in the conduction band. Therefore, it is unclear yet what the origin is of the defect levels that appear at the InP/oxide interface. Although no experimental nor simulated evidence is present to assign the defect levels to a chemical bonding state, we considered minimizing the structural defects at the InP surface by a gas phase HCl etch prior to the ALD deposition. The idea is that an HCl etch will simultaneously remove the oxides and the structural defects at the top surface.

Experiments and discussion. The surface preparation and high-κ deposition were performed in a Polygon® 8200 cluster [33] without air breaks between these process steps. The Polygon cluster is equipped with and EPSILON™ reactor for the pretreatments and an ALCVD™ PULSAR® 2000, hot wall cross flow reactor. A flow of 10 sccm HCl diluted in 20 slm H_2 was exposed to bulk 2 inch InP wafers. In order to achieve a smooth etch, removal rates and surface roughness were determined at different exposure temperatures. In Figure 7, AFM images are shown of the InP surfaces after the HCl etch at 350 C, 400 C and 450 C. RMS data and etch rates of the corresponding surfaces are shown in Table 1.

The etch rate increases with temperature. Increasing the etch temperature also leads to smoother surfaces. Initially at 300 C, the surface roughens due to the HCl etch. However, by increasing the temperature to 450 C, the surface smoothens and exhibits atomically flat terraces. This is an indication of the high quality of the surface. After the HCl etch, the InP wafer is transferred to the ALD chamber without an air break. Al_2O_3 is deposited by TMA/H_2O at 300 C. Although, we assume that the surface is free of defects, the ALD deposition might introduce new defects. Therefore, S-passivation was also introduced. However, an air break would lead to oxidation and surface defect formation. $(NH_4)_2S$ which is commonly used on III/V is therefore replaced by an exposure to H_2S in the gas phase subsequently to the HCl etch.

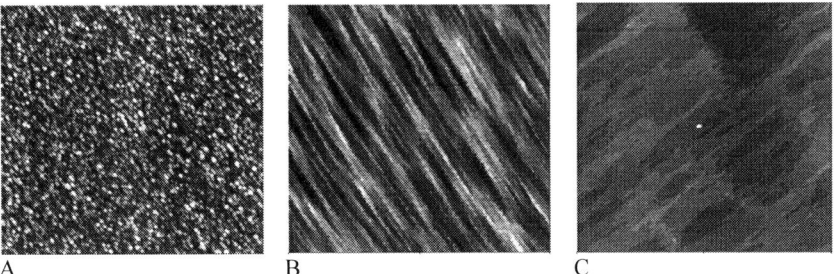

A   B   C

Figure 7: AFM images on a 2x2 μm scale of the InP surfaces etched by a gas phase exposure of HCl at 350 C, 400 C and 450 C for A, B and C respectively.

TABLE I. RMS (nm) values of the InP surface etched by gas phase exposure to HCl at 350 C, 400 C and 450 C and corresponding etch rates.

RMS (nm)	350 C	400 C	450 C
1x1 μm	1.941	0.844	0.205
2x2 μm	2.005	0.938	0.237
Etch rate (nm/min)	<1	9	55

H_2S exposure was performed in the same reactor at 20 Torr. The substrate is exposed to 100 sccm of the diluted H_2S in 20 slm of N_2 gas for 1 min at 320 C. Finally, the substrate temperature was decreased to room temperature in N_2. H_2S did not affect the surface roughness. XPS data of the P2p peak are shown in Figure 8 for different cycles of TMA/H_2O on 2 InP surfaces: A) the InP surface treated with the HCl etch at 450 C for 15s. B) The InP surface treated with the same HCl etch, followed by an H_2S exposure. For both samples, the same trend is present. P-oxides are present at 134 eV and decrease upon increasing the amount of TMA/H_2O cycles. At 20 cycles, the P-oxides peak has completely disappeared. This means that no P-oxides are present for Al_2O_3 layers thicker than 20 cycles. Similar results can be found for the In-oxides. The H_2S treated samples show some residual In-oxides or In-sulfides even for more than 20 cycles of TMA/H_2O. As the binding energy shift between In-O and In-S is very close (energy difference<0.5eV), it is hard to distinguish between the In-oxides and In-sulfides in the case of the H_2S treated InP wafer. The InP wafer treated with HCl only, did not exhibit any In-oxides from 20 cycles on. The interface between the InP and the Al_2O_3 is therefore oxide free. We assume that the starting surface is oxide-free and the ALD deposition does not oxidize the surface and hence does not introduce any additional defects. The presence of oxides at a lower amount of deposition cycles, comes from the oxidation of the InP surface through the gaps during air exposure. As the Al_2O_3 films with a low amount of TMA/H_2O cycles are not closed, re-oxidation can occur through the gaps. This also indicates that layer closure is situated around 20 cycles of TMA/H_2O deposition. In Figure 8 CV spectra are shown for n-InP wafers with a 10 nm Al_2O_3 film with an HCl and HCl/H_2S pretreatment. Pt dots were deposited and a FGA was performed at 400 C for 5 min.

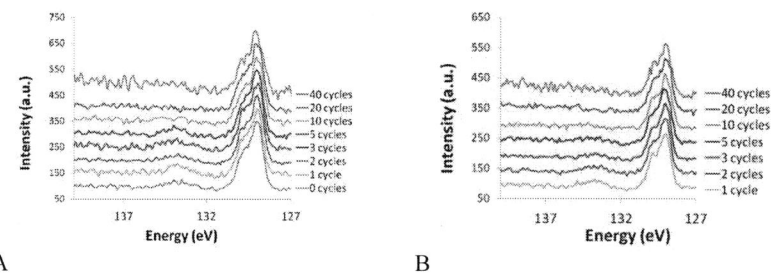

A   B

Figure 8. XPS spectra of the P2p peak measured for different cycles of TMA/H_2O on an InP wafer with a HCl etch at 450 C (A) and on an InP wafer with an HCl etch at 450 C followed by an H_2S exposure.

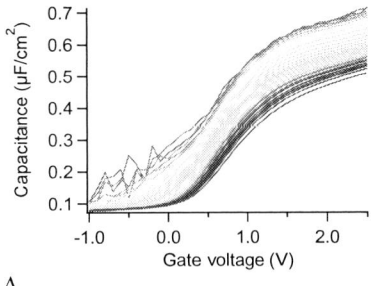 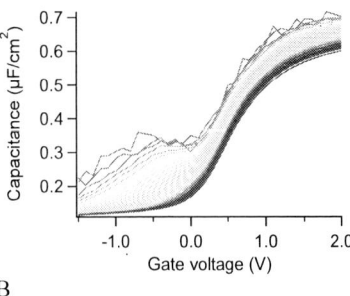

A
                                      B

Figure 9. CV's of the n-InP/Al_2O_3(10 nm)/Pt. A) The InP wafer received an HCl etch at 450 C for 15 s prior to Al_2O_3 deposition. B) The InP wafer received an HCl etch at 450 C for 15 s, followed by an H_2S exposure for 1 min at 320 C prior to Al_2O_3 deposition. The stacks received a FGA anneal at 400 C for 5 min. 30 frequencies were varied logarithmically between 100 Hz (upper line) and 1 MHz (lower line).

It appears that the S-treatment improves the CV of the gate stack: less frequency dispersion is observed at accumulation. However, it is also clear that the HCl etch did not improve the defect states commonly observed at the InP/high-κ interface. Large frequency dispersion is present from depletion to the accumulation region.

We could speculate on the reason of failure of this approach. The first reason could be that the effectiveness of the HCl etch is not as we expected. We do not have an in-situ monitoring system to check the quality of the surface. It might be possible that oxidation of the interface still takes place during ALD deposition but to a minor extent such that the oxides are below the detection limit of XPS. Detection limits of XPS are usually in the range of 10^{14} at/cm^2, while the electrical defect levels are in the range of 10^{11}-10^{13} at/cm^2. There are no physical characterization techniques offering chemical bonding information that have such low detection limits. Therefore, the experimental correlation of the electrically active defects to a chemical bond is very difficult to make.

On the other hand, recent developments in the analysis of the CV data of III/V materials, brought another trapping mechanism to the fore-front: border traps. While interface traps are located at the semiconductor/oxide interface, the border traps are located in the oxide near the interface. Besides the passivation of the interface between III/V and the high-κ oxide, the low conduction band density of states is raising the probability of charge trapping action at accumulation between the oxide border traps and conduction band electrons due to the fact that the surface Fermi level can be biased well into the conduction band [14,15]. Any defects in the oxide will therefore play a role in the final device performance. On n-type InP, the high frequency dispersion in the positive-bias-voltage region is completely dominated by a large border trap distribution [14,15]. Simulations of the CV data of the InP/oxide interface lead to the conclusion that a high D_{it} is present at the valence band edge and relative low D_{it} at the conduction band edge. The border trap distribution peaks at about 0.2 eV above the conduction band edge.

Therefore, focusing only on the interface passivation for III/V materials will not solve all of the passivation issues as the border traps are playing a major role in trapping the mobile carriers. Therefore, any future experiments should also take into account the defects created in the oxide near the interface. However, it is unclear what the chemical/physical nature of these traps is.

Summary

A lot of research has been done on Ge passivation over the last years. Very high quality Ge/GeO_2 gate stacks have been made by high temperature/high pressure growth of GeO_2 on Ge. Although promising defect levels were reported for both nMOS and pMOS applications, the scalability of such a stack is very questionable. We demonstrated that the introduction of S at the Ge/oxide interface can combine low defect levels with low EOT. When Al_2O_3 is deposited by TMA/H_2O, an oxide free interface is obtained with low defect levels towards the valence band edge and therefore suitable for pMOS devices. However, the scalability of this stack is limited by the layer closure of the Al_2O_3 deposition process which is at 20 cycles of TMA/H_2O. By changing the oxidant precursor to O_3, a GeO_xS_y interface is formed that resembles the GeO_x interface. However, the D_{it} levels are superior to the GeO_x interface. This opens new possibilities to scale the Ge gate stack. We have shown that the bi-layer approach does not degrade the Ge/GeO_xS_y interface and can be used as a starting point to further scale the gate stack more aggressively. This approach is currently under investigation.

The passivation of the III/V-oxide interface is even more complex: any oxidation of the surface should be avoided as this introduces stress at the surface. However, reduction of the oxides does not repair the damage done by the oxidation. We investigated the possibility to reduce the damage at an InP surface by a gas phase etch in HCl prior to oxide deposition. Although very high quality surfaces were obtained and oxidation was prevented during the ALD step, the D_{it} did not improve. A possible explanation could be found in the discrepancy between the detection limits of the physical characterization techniques and the electrical analysis techniques. Although, the physical characterization is pointing to an oxide free and high quality surface, the defect levels that are being measured electrically are not in the range of these techniques.

On the other hand, another possible explanation could be the fact that the problem goes beyond the interface. High frequency dispersion measured in accumulation on n-type InP is pointing to border traps. These traps are rather related to the oxide near the interface. This means that future experiments should not only focus on creating the perfect interface but also a defect free oxide.

Acknowledgments

The authors also acknowledge the European Commission for financial support in the DualLogic project no. 214579. We also kindly acknowledge TMEIC (Toshiba-Mitsubishi Electric Systems Corporation) for providing an O_3 generator.

References

1. J. Mitard et al., *IEDM Tech. Dig.*, 1-4, (2008).
2. C. H. Lee, T. Tabata, T. Nishimura, K. Nagashio, K. Kita and A. Toriumi, *Appl. Phys. Exp.*, **2**, 071404 (2009).
3. H. Matsubara, T. Sasada, M. Takenaka, and S. Takagi, *Appl. Phys. Lett.*, **93**, 032104 (2008).
4. F. Bellenger, M. Houssa, A. Delabie, V. Afanasiev, T. Conard, M. Caymax, M.

Meuris, K. De Meyer, and M. M. Heyns, *J. of Electrochem. Soc.*, **155**(2), G33-G38 (2008).

5. T. Sugawara, Y. Oshima, R. Sreenivasan, and P. McIntyre, *Appl. Phys. Lett.*, **90**, 112912 (2007).

6. K. Saraswat, D. Kim, T. Krishnamohan, D. Kuzum, A. K. Okyay, A. Pethe, and H.-Y. Yu, *ECS Trans.*, **16** (10), 3-12, (2008).

7. S. Sioncke, H.C. Lin, G. Brammertz, A. Delabie, T. Conard, A. Franquet, M. Meuris, H. Struyf, S. De Gendt, M. Heyns, C. Fleischmann, K. Temst, A. Vantomme, M. Müller, M. Kolbe, B. Beckhoff, M. Caymax, *J. of Electrochem. Soc.*, **158** (7), H687 (2011).

8. S. Sioncke, H.C. Lin, L. Nyns, G. Brammertz, A. Delabie, T. Conard, A. Franquet, J. Rip, H. Struyf, S. De Gendt, M. Müller, B. Beckhoff, M. Caymax, *J. of Appl. Phys.*, **110** (8), 084907, (2011).

9. R. Zhang, T. Iwasaki, N. Takoka, M. Takenaka, S. Takagi, *Microelectron. Eng.*, **88**, 1533-1536, (2011).

10. R. Zhang, T. Iwasaki, N. Taoka, M. Takenaka, and S. Takagi, *VLSI Techn. Dig.*, 56-57, (2011).

11. R. Zhang, N. Taoka, P.-C. Huang, M. Takenaka, and S. Takagi, *IEDM Tech. Dig.*, 642-644, (2011).

12. G. Brammertz, H.C. Lin, K. Martens, A. Alian, C. Merckling, J. Penaud, D. Kohen, W.-E. Wang, S. Sioncke, A. Delabie, M. Meuris, M. Caymax, M. Heyns, *ECS Trans.*, **19** (5), 375-386 (2009).

13. M. Scarrozza, G. Pourtois, M. Houssa, M. Caymax, A. Stesmans, M. Meuris, M.M. Heyns, *Microelectron. Eng.*, **86**, 1747-1750 (2009).

14. G. Brammertz, H.C. Lin, M. Caymax, M. Meuris, M. Heyns and M. Paslack, *Appl. Phys. Lett.*, **95** (20), 202109-1-202109-3 (2009).

15. G. Brammertz, A. Ali, H.-C. Lin, M. Mueris, M. Caymax, W.-E. Wang, *IEEE Trans. Electron Devices*, **58** (11), 3890-3897, (2011).

16. M. M. Frank, H. Shang, S. Rivillon, F. Amy, C.-L. Hsueh, V. K. Paruchuri, R. T. Mo, M. Copel, E. P. Gusev, M. A. Gribelyuk, and Y. J. Chabal, *Solid State Phenomena*, **103-104**, 3 (2005).

17. M. Houssa, G. Pourtois, M. Caymax, M. Meuris, M. M. Heyns, V.V. Afanasiev, and A. Stesmans, *Appl. Phys. Lett.*, **93**, 161909 (2008).

18. A. Toriumi, T. Tabata, C.-H. Lee, T. Nishimura, K. Kita, K. Nagashio, *Microelectron. Eng.*, **86**, 1571-1576 (2009).

19. A. Toriumi, C.H. Lee, T. Nishimura, S. K. Wang, K. Kita, and K. Nagashio, *ECS Trans.*, **35**(3), 443-456 (2011).

20. R. Ota, M. Kunungi, *J. of Phys. and Chem. of Solids*, **38** (1), 9-13, 1977.

21. A.S. Tverjanovich, *Glass Phys. and Chem.*, **29** (6), 532-536, 2003.

22. C. Fleischmann, S. Sioncke, S. Couet, K. Schouteden, B. Beckhoff, M. Müller, P. Hönicke, M. Kolbe, C. Van Haesendonck, M. Meuris, K. Temst, and A. Vantomme, *J. Electrochem. Soc.*, **158** (5), H589-H594 (2011).

23. A. Delabie, S. Sioncke, J. Rip, S. Van Elshocht, M. Caymax, G. Pourtois, and K. Pierloot, *J. Phys. Chem. C*, **115**, 17523 (2011).

24. A. Delabie, S. Sioncke, J. Rip, S. Van Elshocht, G. Pourtois, M. Mueller, B. Beckhoff, K. Pierloot, J. of Vac. Science and Technol. A, **30** (1), 01A127, (2012).

25. M. Caymax, G. Eneman, F. Bellenger, C. Merckling, A. Delabie, G. Wang, R. Loo, E. Simoen, J. Mitard, B. De Jaeger, G. Hellings, K. De Meyer, M. Meuris, and M. Heyns, in *2009 IEEE International Electron Devices Meeting (IEDM)*, 461 (2009).

26. E. H. Nicolian and J. R. Brews, MOS Phys. and Tech., p. 215-216, Wiley Interscience, New York (2003).

27. W. Wang, C.L. Hinkle, E.M. Vogel, K. Cho, R.M. Wallace, *Microelectron. Eng.*, **88**, 1061-1065 (2011).

28. A.P. Kirk, M. Milojevic, J. Kim and R.M. Wallace, *Appl. Phys. Lett.*, **96**, 202905 (2010).

29. M. Milojevic, F.S. Aguirre-Tostado, C.L. Hinkle, H.C. Kim, E. M. Vogel, *Appl. Phys. Lett.*, **93**, 202902, (2008).

30. M. Tallarida, C. Adelmann, A. Delabie, S. Van Elshocht, M. Caymax, Applied Physics, Letters, 99, 042906 (2011).

31. H.-P. Komsa, A. Pasquerello, *Microelectron. Eng.*, **88**, 1436-1439 (2011).

32. J. Robertson, L. Lin, *Microelectron. Eng.*, **88**, 1440-1443, (2011).

33. Polygon, Epsilon, ALCVD, and PULSAR are trademarks of ASM international, The Netherlands.

Investigations on thermal stress relief mechanism using air-gapped SiO$_2$ nanotemplates during epitaxial growth of Ge on Si and corresponding hole mobility improvement

Swapnadip Ghosh[1], Darin Leonhardt[2], and Sang M. Han[1,2]

(1) Department of Electrical and Computer Engineering, University of New Mexico, Albuquerque, New Mexico 87131, USA
(2) Department of Chemical and Nuclear Engineering, University of New Mexico, Albuquerque, New Mexico 87131, USA

We demonstrate the implementation of air-gapped SiO$_2$ nanotemplates embedded in epitaxially grown Ge on Si for relieving stress caused by the thermal expansion coefficient mismatch between Ge and Si. The air-gap is formed around the SiO$_2$ template during growth and eventual coalescence of adjacent Ge islands merging over the template. The stress map obtained from finite element modeling corroborates the experimental observation, suggesting that the thermal stress can be reduced nearly by half. The templates also filter threading dislocations propagating from the underlying Ge-Si interface, reducing the defect density from 9.8×10^8 to 4.4×10^7 cm^{-2} in the demonstration case. We further investigate the influence of threading dislocation density on the effective hole mobility in undoped Ge between substrates grown with the template and without the template. Using the Hall mobility measurements, we have obtained a peak effective hole mobility of 925 cm^2/V-s at room temperature for Ge grown with the template, compared to 297 cm^2/V-s for Ge grown without the template.

Introduction:

Low-defect-density, relaxed Ge, epitaxially grown on Si with a significantly greater hole mobility than Si, presents a promising platform for high mobility p-MOS devices.[1,2] Growing low-dislocation-density Ge on Si (GoS) and subsequently integrating III-V layers presents two significant engineering challenges: lattice and thermal expansion coefficient mismatch. The materials engineering solutions to circumvent the lattice mismatch include post-growth annealing[3], graded buffer layers[4], selective epitaxial overgrowth (SEG)[5], and aspect ratio trapping (ART)[6]. The ART technique, in particular, utilizes high-aspect-ratio holes or trenches etched through dielectric films to trap dislocations, greatly reducing the dislocation density. A noteworthy advantage of ART technique is that it avoids the thick buffer and high thermal budget typical of other heteroepitaxial techniques, making it more suitable for integration with Si complementary metal oxide semiconductor (CMOS) process. However, one shortcoming of ART is that it has been demonstrated to be effective only for small holes or narrow strips with dimensions less than 1 μm.[7]

Experimental Results:

We focus on the use of air-gapped SiO_2-based templates with nanoscale channels placed on the epilayer of GoS followed by Ge SEG. Figure 1 shows cross-sectional transmission electron microscopy (x-TEM) images of the structure containing the SiO_2 nanotemplate. In Figure 1, the estimated threading dislocation density (TDD) below the oxide template is 9×10^8 cm^{-2}. We note that this relatively high TDD in the lower Ge epilayer is chosen as the baseline solely for the purpose of demonstrating of TD filtering. Most TDs are blocked by the oxide walls and do not propagate into the upper SEG Ge layer. Figure 1 also shows that voids (or air gaps) form around the sidewalls and top of the oxide template during Ge SEG at 923K. The template simultaneously filters threading dislocations (TDs) propagating from the Ge-Si interface and relieves the film stress caused by the thermal expansion coefficient mismatch between Ge and Si. The templates also filter threading dislocations propagating from the underlying Ge-Si interface, reducing the density from 9.8×10^8 to 1.6×10^7 cm^{-2}. We analyze the effectiveness of the template in filtering TDs in the lower Ge layer.

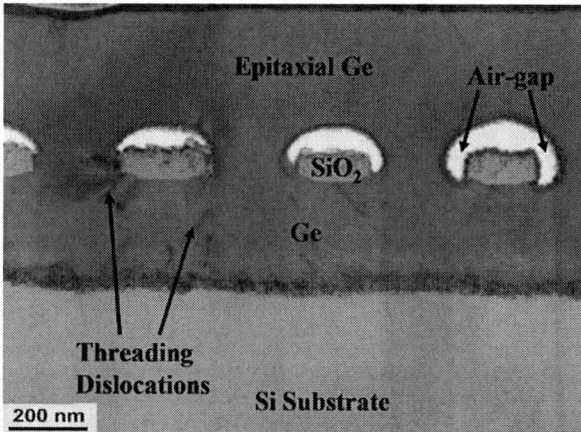

Figure 1. TEM image of Ge forming voids around the SiO_2 template, while epitaxially growing on GoS. The template filters threading dislocations.

Analysis:

We also examine the existence and potential causes of defects stemming from the coalescence of adjacent Ge growing out of the template channels over the air-gapped SiO_2 template. We consider possible mechanisms responsible for the twin/SF formation. Coalescence defects may form due to translation misalignment. Thermal stress in the Ge epilayer is caused by the difference in thermal expansion coefficient (TEC) of Si, Ge, and SiO_2. Basically, the thermal stress induces a varying Ge lattice constant in the underlying GoS, especially adjacent to SiO_2, thereby leading to translational mismatch during subsequent lateral overgrowth. Next, we use FEM to investigate the effects of SiO_2 template and Ge morphology on the thermal stress in the samples. Figure 2 corresponds to the structure shown in Figure 1, in which Ge forms a void over SiO_2 nanotemplate. The high stress regions indicated by arrows are where twin defects are observed in Figure

1. The comparison amongst stress maps of Ge grown with and without the air-gapped template also strongly suggests that the thermal stress can be reduced nearly by half by inserting the air-gapped template in the Ge epilayer.

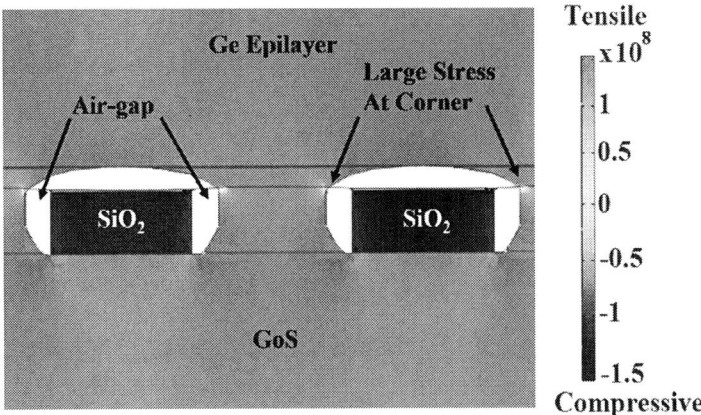

Figure 2. FEM stress simulation of Ge with air-gap around the SiO_2 template, while epitaxially grown on GoS

Lastly, we have experimentally investigated the influence of TDD on the effective hole motility in Ge grown with and without the air-gapped template on GoS. Using the Hall measurements, we have obtained a peak effective hole mobility of 925 cm^2/V-s at room temperature for Ge grown with the air-gapped template, compared to 297 cm^2/V-s for Ge grown directly on Si without the template.

In summary, an air-gapped SiO_2 nanotemplate structure is considered to reduce the thermal stress caused by TEC mismatch between Ge and Si and to simultaneously filter the TDs. A FEM stress modeling further corroborates this stress relief mechanism. Based on the effective mobility values, the epitaxially grown Ge layer using air-gapped SiO_2-based templates may potentially prove useful for wafer-scale integration of high mobility p-MOS transistors on Si.

Acknowledgments
The above material is based upon work supported by, or in part by, the National Science Foundation (DMR-0907112 and CMMI-1068970). The authors also acknowledge the use of TEM facility at the University of New Mexico.

References
[1] D. P. Brunco, B. De Jaeger, G. Eneman, J. Mitard, G. Hellings, A. Satta, V. Terzieva, L. Souriau, F. E. Leys, G. Pourtois, M. Houssa, G. Winderickx, E. Vrancken, S. Sioncke, K. Opsomer, G. Nicholas, M. Caymax, A. Stesmans, J. Van Steenbergen, P. W. Mertens, M. Meuris, and M. M. Heynsa, "Germanium MOSFET Devices: Advances in Materials Understanding, Process Development, and Electrical Performance," *J. Electrochem. Soc.* **155,** H552-H561 (2008).

[2]   H.-Y. Yu, M. Ishibashi, J.-H. Park, M. Kobayashi, and K. C. Saraswat, "p-Channel Ge MOSFET by Selectively Heteroepitaxially Grown Ge on Si," *IEEE Electron Device Lett.* **30,** 675-677 (2009).

[3]   D. R. L. H. -C. Luan, K. K. Lee, K. M. Chen, K. Wada, and L. C. Kimerling," *Appl. Phys. Lett.* **75,** 2909 (1999).

[4]   S. B. S. M. T. Currie, T. A. Langdo, C. W. Leitz, and E. A. Fitzgerald," *Appl. Phys. Lett.* **72,** 1718 (1998).

[5]   Y.-B. J. Q. Li, H. Xu, S. Hersee, and S. M. Han," *Appl. Phys. Lett.* **85,** 1928 (2004).

[6]   M. C. J. -S. Park, J. M. Hydrick, J. Bai, M. Carroll, J. G. Fiorenza, and A. Lochtefeld,," *J. Electrochem. Society* **156(4),** H249 (2009).

[7]   M. C. J. -S. Park, J. M. Hydrick, J. Bai, J. -T. Li, Z. Cheng, M. Carroll, J. G. Fiorenza, and A. Lochtefeld," *Electrochem. Solid-state. Lett.* **12(4),** H142 (2009).

Integration of InGaAs Channel n-MOS Devices on 200mm Si Wafers Using the Aspect-Ratio-Trapping Technique

Niamh Waldron, Gang Wang[1], Ngoc Duy Nguyen[2], Tommaso Orzali, Clement Merckling, Guy Brammertz, Patrick Ong, Gillis Winderickx, Geert Hellings, Geert Eneman, Matty Caymax, Marc Meuris, Naoto Horiguchi and Aaron Thean

Imec, Kapeldreef 75, B-3001 Leuven, Belgium
currently at [1]MEMC, Electronic Materials Inc.
[2] Université de Liège, Institut de Physique B5a, B-4000 Liège, Belgium

We report on the fabrication on InGaAs/InP implant free quantum well (IFQW) n-MOSFET devices on 200mm wafers in a Si CMOS processing environment. The starting virtual InP substrates were prepared by means of the aspect-ratio-trapping technique. Post CMP these substrate resulted in a planar substrate with a rms roughness of 0.32 nm. After channel and gate processing source drain regions were formed by the selective epitaxial growth of Si doped InGaAs. Contact to the source/drain regions was made by a standard W-plug/metal 1 process. The contact resistance was estimated to be on the order of $7x10^{-7}$ $\Omega.cm^2$. Fully processed devices clearly showed gate modulation albeit on top of high levels of source to drain leakage. The source of this leakage was determined to be the result of the unintentional background doping of the InP buffer layer. Simulations show that the inclusion of the p-InAlAs between the InP and InGaAs can effectively suppress this leakage. This development is a significant step towards the integration of InGaAs based devices on a standard CMOS platform.

Introduction

As CMOS continues to scales to more advanced technological nodes, new higher mobility channel materials will have to be introduced as an alternative to Si in order to meet power and performance requirements [1]. III-V materials and Ge have emerged as an attractive option for nMOS and pMOS respectively. Aside from the difficulties of integrating these new materials in a CMOS flow a critical requirement for their introduction at the ULSI level is that they can be integrated on large size Si wafers. This is a considerable technological challenge as the lattice mismatch between Ge, $In_{0.53}Ga_{0.48}As$ and Si is 4% and 8% respectively. $In_{0.53}Ga_{0.48}As$ devices have been fabricated on 200mm Si wafers by means of growing a thick InAlAs strain relaxed buffer (SRB) layer by MBE [2] but this method is limited by the slow-throughput MBE process and the complexity of ultimately integrating Ge devices on such a pseudo blanket InGaAs wafer. Another approach is to use direct wafer bonding (DWB) to bond a III-V layer to a Ge [3] or Ge-on-Si wafer but this will require 300mm or 450mm donor III-V wafers thereby increasing the cost of the starting material significantly. In contrast, the aspect-ratio-trapping (ART) [4] technique offers an elegant and cost effective solution to integrating Ge and III-V materials on the same Si wafer. Starting with an STI template, the Si can be recessed and Ge and III-V alternately grown. It has previously been shown that high performance Ge devices can be fabricated from such an STI template [5]. For III-V materials the use of narrow width trenches allows for defects, which result from the lattice mismatch, to be trapped at the STI sidewall. In this work we present the first integration and electrical results of InGaAs devices fabricated on 200mm wafers using the aspect-ratio-trapping technique.

Transistor Process Flow

The main purpose of this work is to investigate the integration of III-V devices in a Si processing environment using virtual substrates generated by the ART process. To this purpose, the test vehicle chosen was that of a planar InGaAs/InP quantum well device. The integration modules developed can be easily ported to a subsequent finfet flow. A schematic of the process flow used to fabricate the InGaAs channel transistors is shown in Fig 1. The flow begins with defining a template for the InP growth in STI wafers. The height of the STI trenches is 300 nm and the width of the active regions is restricted to 100 nm to 200 nm. This gives aspect ratios from 1.5 to 3 which are necessary for the ART process. After the STI process the Si in the active regions is etched in-situ by HCl vapour and a Ge seed layer deposited on which the nominally undoped InP buffer layer is subsequently grown. The InP is overgrown to a level of 300 nm above the level of the STI trenches and the surface then planarized with a CMP step. Post the InP CMP a 15nm InGaAs channel layer is grown. The gate stack is then deposited which comprises of 10 nm Al_2O_3 as the gate dielectric, 10 nm TiN as the gate metal and 80nm SiO which acts as a hardmask. This gatestack is then patterned. The oxide and TiN are dry-etched and the Al_2O_3 is removed by a wetetch. Narrow SiN only spacers are then formed and Si doped InGaAs is selectively deposited to form the source/drain regions. The oxide hardmask is removed from gate pads so that contact can be made to the TiN gate metal. Next the pre metal dielectric (PMD) stack is deposited and planarized. Contacts holes are opened over the InGaAs S/D regions and gate pad. A Ti/TiN liner layer is deposited followed by W which is also planarized. The processing is then completed by using a standard one level Cu metal damascene backend.

Integration Module Development

InP Epitaxy

Starting with the STI template the Si in the active regions is etched in-situ in an CVD reactor by means of a HCl vapour etch thus creating trenches for the subsequent InP growth. Immediately after the Si etch out a thin 50nm Ge seed layer is grown in-situ in the trenches. The Ge layer acts an intermediate buffer layer between the Si and InP. The formation of double steps which can suppress the formation of anti phase boundary (APB) defects can also be more easily promoted on a Ge rather than Si surface [6]. After the thin Ge seed layer growth, the selective area growth (SAG) of the nominally undoped InP buffer layer is carried out using an Aixtron close-coupled showerhead metal-organic vapor phase epitaxy (MOVPE) systems. Trimethylindium (TMIn) was used as the group-III precursor. Tertiarybutylarsine (TBAs) and tertiarybutylphosphine (TBP) were

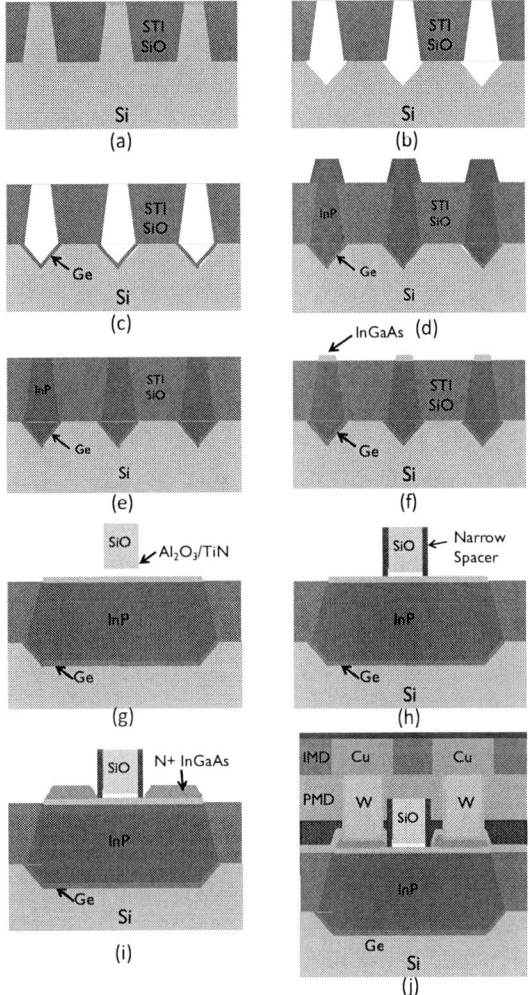

Fig 1: [Color online] Process flow of InGaAs IFQW device. The process starts with an STI template (a). The Si in the trenches is selectively etched (b) and a Ge seed layer deposited (c) on which InP is grown (d). The InP is CMP'd to form a planar surface (e) and the InGaAs channel grown (f). The gate stack is deposited and etched (g) and narrow spacers formed (h). InGaAs N+ S/D regions are grown and the process completed with standard W-plug Metal 1 processing.

employed as the group-V precursors. Before the InP growth, a pre-epi bake at 720 °C and 450 Torr was carried out to remove the Ge native oxide and to promote the formation of double steps. TBAs was introduced during the pre-epi bake to form an As terminated Ge surface that facilitates the InP nucleation [7]. After this thermal treatment the Ge layer is fully relaxed, as confirmed by X-ray diffraction analysis (XRD). Following this bake, the temperature was ramped down to 420 °C for the selective growth of a 30 nm thick InP nucleation or seed layer. Next, the temperature was ramped to 640 °C for the bulk InP SAG. The InP layer was grown to a height of about 300 nm above the level of the STI. More details of the InP growth conditions are described elsewhere [6- 8]. Typical results of the InP growth are shown in Fig. 2. In the direction across the width of the trench the defect density at the top of the trench is low and improves with the aspect ratio of the trench however defects are not fully suppressed in the direction along the length of the trench giving an average defect density on the order of 10^9 cm^{-2}. By using an STI pattern with a more uniform layout of the trenches the defect density can be reduced by an order of magnitude.

InP CMP

The InP overgrowth step is needed as the growth rate of the InP in different width trenches varies because of loading effects. The InP also grows with <111> facets creating a non-planar surface unsuitable for transistor integration. Therefore a CMP step is needed to planarize the surface before the InGaAs channel deposition [9]. Two different slurries were compared for the CMP process. Slurry A is acidic based and normally used for the CMP of W layers. Slurry B is basic and normally used for the CMP of polysilicon layers. It was found that the acidic based slurry gave better process results (Fig. 3). A cross section of a trench planarized using the acidic slurry is shown in Fig. 4(a). The recess between the InP and the STI SiO layer was measured by AFM to be about 8nm (Fig. 4(b)). AFM analysis of the trenches shows the surface roughness of the InP layer post CMP to be 0.32 nm (rms) (Fig.4 (c)).

InGaAs Channel Growth

Following the CMP of the InP surface the InGaAs channel layer is selectively deposited again using the Aixtron-CRIUS MOCVD reactor. For the InGaAs layer growth we used trimethylindium (TMIn) and trimethylgallium (TMGa) as group-III elements precursors and tertiarybutylarsine (TBAs) for group-V element. Any native oxide formed on the InP layer is first desorbed by means a bake carried out at 500 °C. The InGaAs layer was grown at 520 °C by flowing 16 μmol/min of TMGa, 19 μmol/min of TMIn and 2600 μmol/min of TBAs; the ramping up of the sample temperature was performed under TBP flow (2000 μmol/min) and the TBP to TBAs switch was done a number of seconds before allowing the group III precursors in the reactor. The InGaAs layer is nominally undoped.
Cross sectional TEM along the [-110] direction perpendicular to the trenches (Fig. 5(a)) shows a smooth, 15 nm thick InGaAs channel with a sharp InP (001)/InGaAs interface, characterized by a certain degree of roughness that nevertheless seems not to affect the quality of the channel, which exhibits a relatively low density of defects along this direction. {111} and {113} facets are present at the edges of the InGaAs layer. A SEM tilted image of the grown layer is shown in Fig. 5(c): the InGaAs channel appears smooth but a number of 'chops' are visible in the layer. Bright field cross sectional TEM of the III-V layer stack along the [110] direction parallel to the trenches indicates the origin of

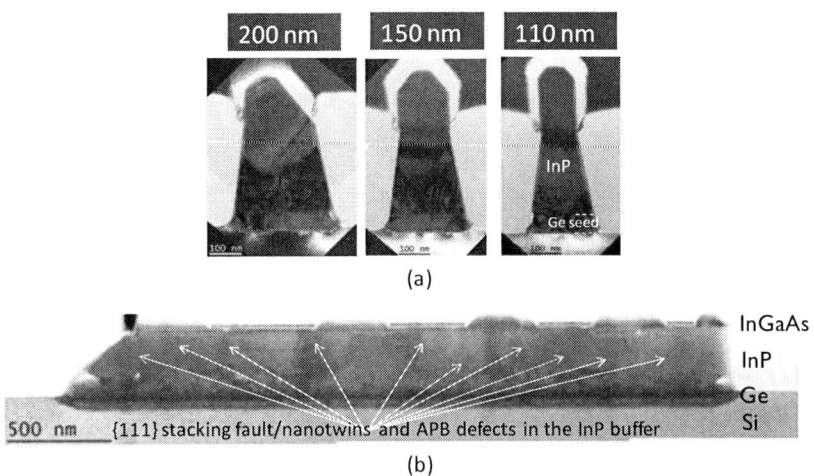

(a)

(b)

Fig 2: a) Bright field cross-sectional TEM images of different width trenches shows that the quality of the InP increases as the aspect ratio increases b) cross-sectional TEM image along the trench reveals a defect density on the order of 10^9 cm^{-2}

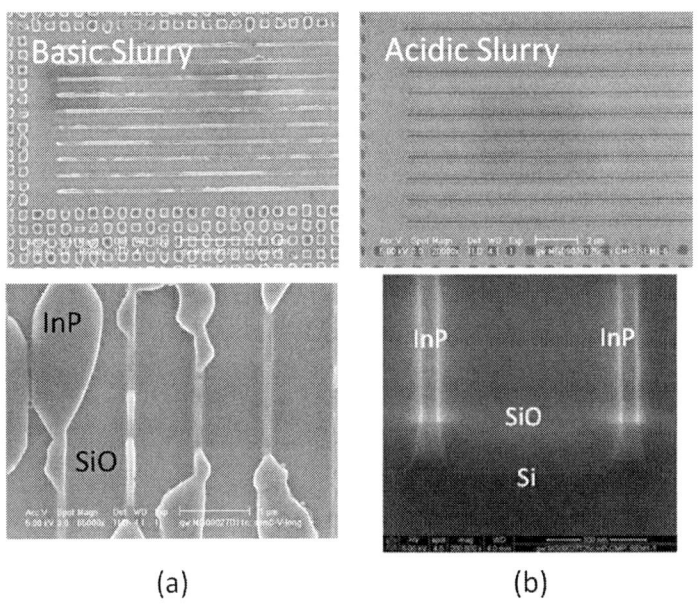

(a)   (b)

Fig 3: a) Top view SEM images of InP in STI trenches post CMP with basic slurry and b) top view and cross sectional images of the InP post CMP with acidic slurry.

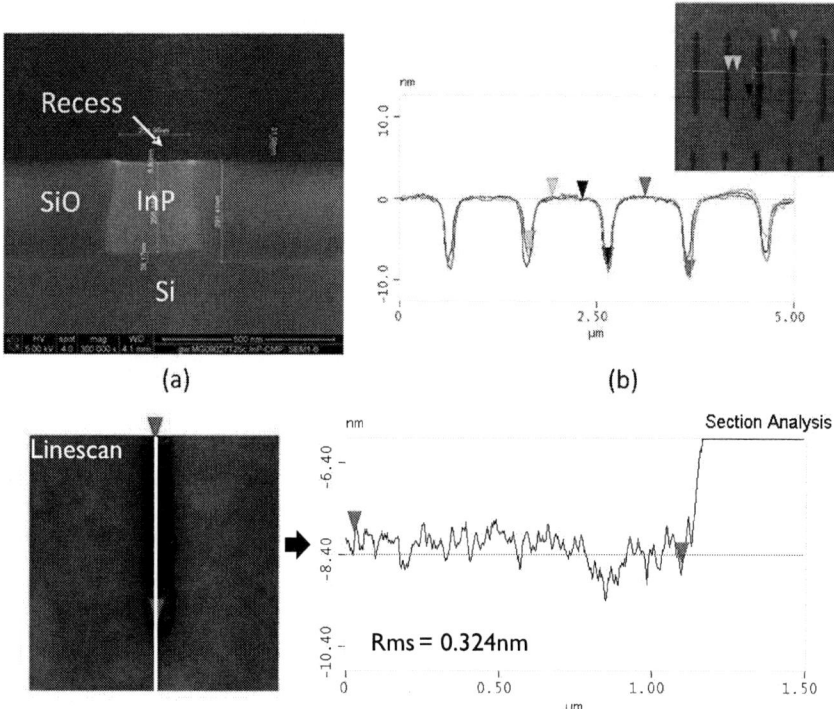

(a)

(b)

Fig 4: a) [Color online] Cross-sectional XSEM of CMP'd InP. b) The depth of the InP recess inside the STI trench was measured by AFM to be 8nm (c) AFM analysis of the CMP'd InP in the STI trench

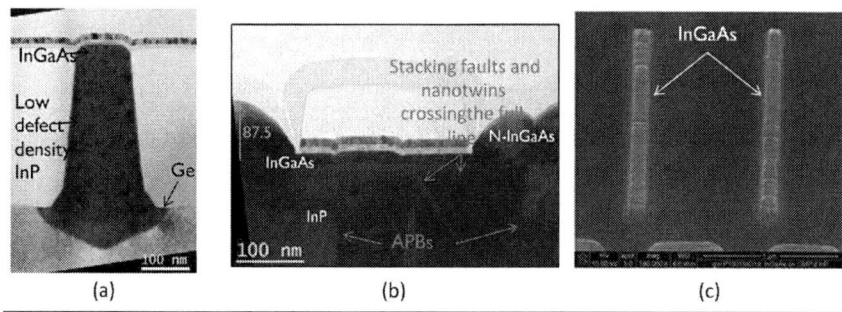

(a)   (b)   (c)

Fig 5: (a) TEM of InGaAs channel growth on the CMP'd InP shows crytalline layer is achieved. (b) Along the length of the trench it can be seen that defects from the InP buffer layer propagate through the InGaAs layer. (c) Top-down SEM view of the InGaAs layer shows 'chops' across the width of the trench which are due to defects in the InP layer as seen in (b).

these features (Fig. 5(b)): grooves form whenever a {111} oriented defect (nanotwin, twin, stacking fault) or a {110} APB originating at the InP/Ge interface cross the InP/InGaAs interface and reach the surface. Therefore, further defect reduction in the InP buffer layer is needed before improvement in the morphology of the InGaAs channel can be realized.

Gate Stack

For the passivation of InGaAs surfaces we have previously shown that the use of sulfur is effective in creating an unpinned surface with Dit levels on the order of 10^{12} eV^{-1}cm^{-2} [10]. However, for these integrated n-MOS devices, the sulfur pre-treatment could not be applied as the aqueous $(NH_4)S_2$ solution used has a too high metallic contamination level to be used in a VLSI fabrication line. Instead a dilute HCl clean was applied immediately before the high-κ deposition of 10 nm Al_2O_3 by ALD. To apply sulfur treatment in the future an integrated gaseous S vapour solution will be needed [11]. A TEM of the resultant InGaAs/Al_2O_3 interface is shown in Fig. 6. Some {111} defects can be seen in the InGaAs layer but the Al_2O_3/InGaAs interface is reasonably smooth. The gate stack patterned by the dry etch of the oxide hard mask and TiN gate metal layer. The dry etch stops on the Al_2O_3 which is then removed by wet etch in a 0.03M HF/10% HCl mixture which is selective to InGaAs.

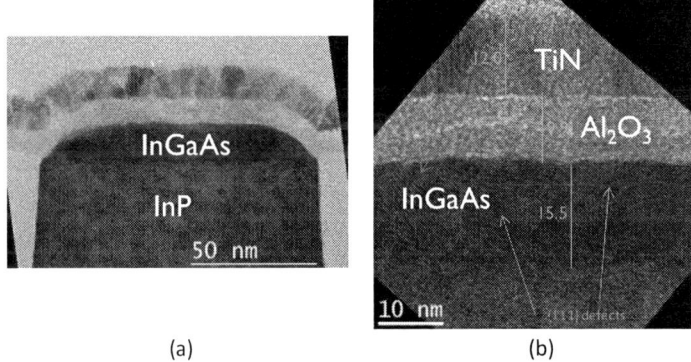

(a)   (b)

Fig 6: TEM of InGaAs/Al_2O_3 interface (a). Some {111} defects can be seen in the InGaAs layer (b) but the Al_2O_3/InGaAs interface is reasonably smooth.

N+ InGaAs Source/Drain Processing

After gate patterning, spacers are deposited. For the IFQW architecture a narrow spacer <10 nm is required to prevent an excessive access resistance [12,13]. A SiN only spacer process was developed to achieve such a thin layer. After spacer formation the wafers are cleaned in a 0.03M HF/10% HCl solution to help remove any residues or damage from the etch. N+ Si doped InGaAs is then selectively grown to form the source/drain regions. The doping of the InGaAs was estimated from TLM measurements to be on the order of mid to high 10^{18}cm^{-3} levels. TEM analysis of the transistor show that the final spacer thickness achieved was 6nm (Fig. 7). The InGaAs S/D growth was selective to the SiN and showed no encroachment under the gate.

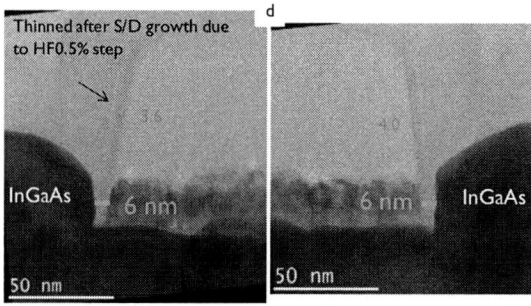

Fig 7: TEM of SiN spacer shows that a 6nm width was achieved. No encroachment of the N+ InGaAs layer underneath the spacer was observed.

Contacts

To form contacts to the InGaAs S/Ds a standard CMOS W-plug scheme was employed. First a PMD stack is deposited and planarized by CMP. Next contact holes are etched. The contacts in this case were drawn as 250 nm x 2 μm. The large contacts were used to minimize any impact of contact resistance on transistor characteristics. After processing the contact size at the bottom of the hole was about 50nm (Fig 8(a)). The contact etch was not selective to the InGaAs even though a CF4 based chemistry was used which should in principle be selective to InGaAs so it is possible that physical etching of the layer is taking place. The etch was timed to stop in the InGaAs layer. A 20 nm Ti/10 nm TiN liner layer was then deposited. Before the Ti deposition an in-situ soft sputter Ar etch was used to remove any native oxide. The contact holes were then filled with W followed by a planarization step. The devices are then completed by using a standard one level Cu metal damascene backend. The highest temperature process in this backend is 400 ˚C.

Physical analysis of the contact by EELS and EDS show that there is no intermixing of the InGaAs and Ti liner layer (Fig. 8(b)). This is an important result as it shows that the non-alloyed ohmic between the InGaAs and Ti can withstand the thermal budget of a standard VLSI CMOS backend. However, the analysis did show that there is still an interfacial oxide between the Ti and InGaAs indicating that the Ar soft sputter etch could be further optimized.

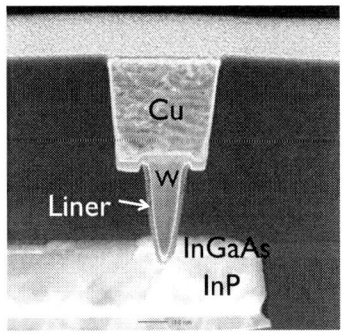

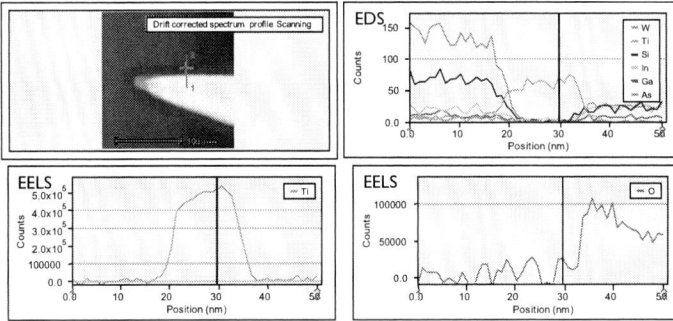

Fig 8: [Color online] TEM of the W-plug contact to the N+ InGaAs S/D layer. The contact etch was not selective to the InGaAs layer. The liner layer consists of Ti/TiN. The width of the contact in the InGaAs layer is about 50 nm. (b) Physical analysis of the contact/InGaAs. EELS and EDS show that there is no intermixing of the Ti from the liner layer and the InGaAs demonstrating that the thermal budget of the backend is low enough. However, oxygen was found at the Ti/InGaAs interface.

Device Results and Discussion

Id-Vgs characteristics

Transistors were fabricated with the process flow described above with the simple InGaAs/InP IFQW architecture. Id-Vgs characteristics of a 130 nm gate length device are shown in Fig. 9. The transistor clearly shows gate modulation but on top of very high levels of source to drain leakage. Physical analysis of the InP grown in the STI implies suggests that this source to drain leakage is a result of the InP layer being conductive. Photoluminesence (PL) measurements of the selectively grown InP are compared to that of a reference InP layer grown on a bulk 2" wafer show a shift in the peak of the response (Fig. 10(a)). This shift corresponds to a n-type doping level on the order of $\sim 10^{17}$ cm^{-3}. The conductive nature of the InP buffer layer has also been confirmed by 2D scanning spreading resistance microscopy (SSRM) analysis on a fully processed wafer (Fig. 10(b)). The SSRM analysis also showed that the Ge seed layer and underlying Si are also highly conductive. This has been shown by SIMS measurement to be a result of Phosphorous doping. The thermal budget of the InP buffer layer growth is 640 °C which makes this level of P diffusion quite surprising. The physical mechanism behind the P diffusion into the Ge and Si is under investigation. The source of the n-type doping in the InP is believed to be background C incorporation from the metal organic precursors into the layer during growth.

For the final device structure including an InAlAs layer between the InP and the InGaAs channel is preferred as the conduction band offset between InAlAs/InGaAs (~0.5 eV) is higher that that between InP/InGaAs (~0.2 eV) [14]. Simulations of this structure show that if the InAlAs is doped p-type and is thick enough the source to drain leakage can be effectively suppressed (Fig. 11). The transport models in the simulations are not calibrated but the results demonstrate the beneficial effects of the p-InAlAs layer on the leakage current.

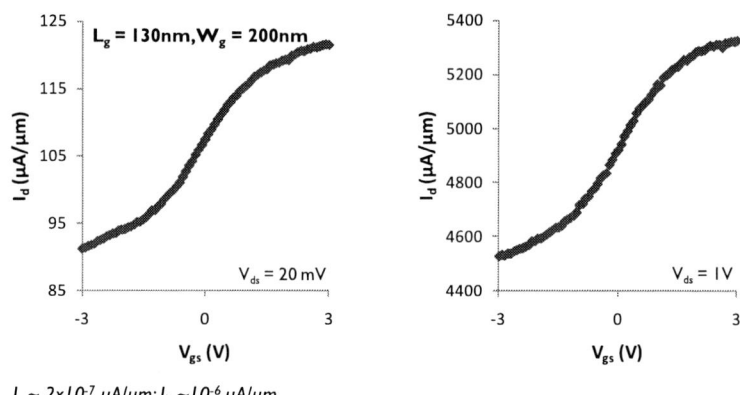

$I_g \sim 2{\times}10^{-7}\ \mu A/\mu m;\ I_b \sim 10^{-6}\ \mu A/\mu m$

Fig 9: Id-Vgs characteristics of fabricated transistors show gate modulation on top of very high levels of source-drain leakage

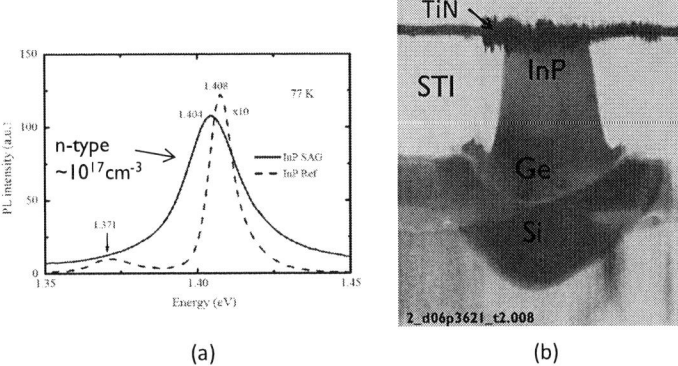

(a)   (b)

Fig. 10: [Color online] (a) PL measurements of InP grown in STI trenches shows a shift in the peak of the response compared to a blanket InP layer implying a n-type on the order of 10^{17} cm^{-3}. (b) The conductive nature of the InP layer in the trenches is confirmed by SSRM measurements. It is also noted that the Ge and Si regions under the trench are also highly conductive. This has been confirmed by SIMS to be P doping.

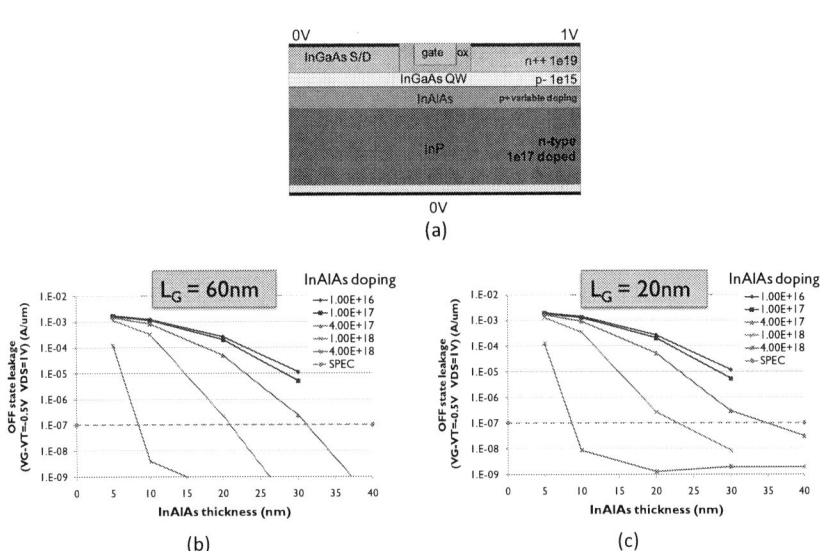

Fig 11: (a) [Color online] Schematic of the structure used to simulate the effect of including a p-type InAlAs layer in the quantum well structure. (b) and (c) Off state leakage for 60 nm and 20 nm gate length devices as a function of the InAlAs layer thickness and doping.

Contact Resistance

TLM measurements of the N+ InGaAs layer yielded an estimate of the specific contact resistivity on the order of 7×10^{-7} $\Omega.cm^2$ (Fig. 12). This is a worse case estimate as we take the whole area of the contact into account. The ITRS roadmap indicates that contact resistances on the order of 5×10^{-8} $\Omega.cm^{-2}$ are required for advanced nodes [15]. In order to achieve a lower contact resistance the InGaAs doping can be increased [16] and/or an InAs contact layer can be used [17].

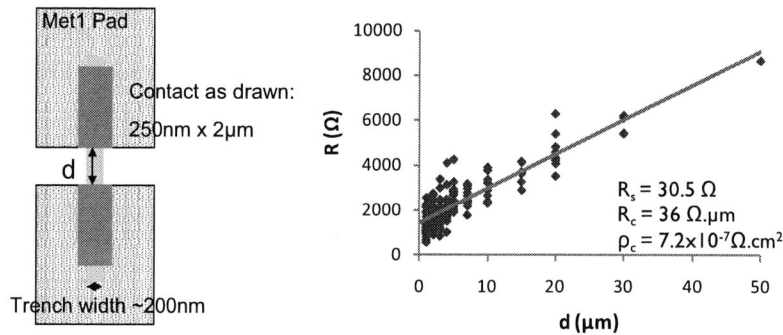

Fig 12: Taking a contact area of 50nm x 2μm, an ρ_c of 7.2×10^{-7} $\Omega.cm^2$ was extracted from TLM measurements.

Conclusions

We have demonstrated the first fully integrated InGaAs n-MOSFET devices processed using the ART technique. The required integration modules to fabricate an InGaAs channel IFQW device were developed in a 200mm Si CMOS processing line. Planar patterned virtual III-V substrates were generated by the growth of an InP buffer layer in a standard Si STI template. The InP layer was subsequently polished giving a rms roughness of 0.32nm. Despite the ART effect along the width of the trench defects still remained in the direction along the length of the trench giving a typical defect density on the order of 10^9 cm^{-2}. These defects are a limiting factor on the morphology of the InGaAs channel layer. Pits in the surface of the channel are observed where dislocations, stacking faults or APB defects reach the surface of the buffer layer. Further optimization of the Ge seed layer and InP buffer growth is required to reduce the level of defects in the channel region. An effective narrow 6nm SiN spacer process was developed for the selective epitaxial growth of Si doped InGaAs source and drain regions. Contact to the N+ source drain regions was achieved through the use of a standard CMOS W-plug process. A worst case contact resistance of 7×10^{-7} Ωcm^{-2} was extracted from TLM test structures. Physical characterization of the contacts shows that there is no intermixing of the InGaAs and Ti liner of the W-plug due to the thermal budget of the Metal 1 module (≤ 400 °C). The doping of the N+ source drain regions was estimated to be mid to high

10^{18} cm^{-3} levels. Further improvement in the R_c can achieved by increasing this doping level and adding an InAs capping layer. Fully processed devices showed gate modulation on top of very high levels of source/drain leakage (~mA/μm). This leakage was confirmed to be the result of the InP buffer layer being conductive. Simulations show that the leakage can be suppressed by the inclusion of a p-InAlAs layer between the InP and InGaAs channel layer.

While these results demonstrate that many challenges still remain for the integrated of high performance InGaAs devices on a Si platform, they are a significant step towards the realization of an ART based solution.

Acknowledgements

The authors acknowledge the European Commission for financial support in the DualLogic project no. 214579. Further, we thank the imec core partners with the IIAP on Logic-DRAM.

References

1. A.Khakifirooz and D.A. Antoniadis, *Trans. Electron. Devices*, **55**, p1401-8, Jun 2008
2. R. Hill, C.Park, J. Barnett, J. Price, J. Huang, N. Goel, W. Loh, J. Oh, C. Smith, P.Kirsch, P. Majhl and R. Jammy, *IEDM Tech. Digest*, 2010
3. M. Yokoyama, S. Kim, R.Zhang, N. Taoka, Y. Urabe, T. Maeda, H. Takagi, T. Yasuda, H. Yamada, O. Ichikawa, N. Fukuhara, M. Hata, M. Sugiyama, Y. Nakano, M. Takenaka and S. Takagi, *VLSI Tech. Digest* 2011
4. E.A. Fitzgerald and N.Chand, *Electron. Mater*, **20**, pp 839-53, Oct 1991
5. J. Mitard, C. Shea, B. DeJaeger, A. Pristera, G. Wang, M. Houssa, G. Eneman, G. Hellings, W.E. Wang, J, Lin, F. Leys, R. Loo, G. Winderickx, E. Vrancken, A. Stesmans, K. DeMeyer, M. Caymax, L. Pantisano, M. Meuris and M. Heyns, *VLSI Tech. Digest*, pp. 82-3, 2009
6. G. Wang, M. Leys, R. Loo, O. Richard, H. Bender, G. Brammertz, N. Waldron. W-E Wang, J. Dekoster, M. Caymax, M. Seefeldt and M. Heyns, *J. Electrochem. Soc.*, **158**(6), H645, 2011
7. G. Wang, M.Leys, N. Nguyen, R. Loo, G. Brammertz, O. Ricahrd, H. Bender, J. Dekoster, M. Meuris, M. Heyns and M. Caymax, *J. Electrochem. Soc.*, **157**(1), H1023 (2010)
8. G. Wang, M. Leys, N. Nguyen, R. Loo, G. Brammertz, O. Richard, H. Bender, J. Dekoster, M. Meuris, M. Heyns and M. Caymax *J. Cryst. Growth*, **315**, pp 32-6 2011
9. P. Ong, L. Witters, N. Waldron and L. Leunissen, *ECS Trans*. Vol. 34, Issue 1, 2011
10. E. O'Connor, B. Brennan, V. Djara, K. Cherkaoui, S. Monaghan, S. B. Newcomb, R. Contreras, M. Milojevic, G. Hughes, M. E. Pemble, R. M. Wallace, and P. K. Hurley. *J. Appl. Phys,* **109**, 024101, 2011

11. A. Alian, G. Brammertz, C. Merckling, A. Firrincieli, W.E. Wang, H.C. Lin, M. Caymax, M. Meuris, K. De Meyer and M. Heyns, *Appl. Phys. Lett.*, **99**,11, pp 112114, Sep 2011

12. G. Hellings, L. Witters, R. Krom, J. Mitard, A. Hikavvy, R. Loo, A. Schulze, G. Eneman, C. Kerner, J. Franco, T. Chiarella, S. Takeoka, J. Tseng, W.E. Wang, W. Vandervorst, P. Absil, S. Biesemans, M. Heyns, K. De Meyer, M. Meuris and T. Hoffmann, *IEDM Tech Digest*, 2010

13. J. Mitard, L. Witters, G. Hellings, R. Krom, J. Franco, G. Eneman, A. Hikavvy, B. Vincent, R. Loo, P. Favia, H. Dekkers, E. Altamirano Sanchez, A. Vanderheyden, D. Vanhaeren, P. Eyben, S. Takeoka, S. Yamaguchi, M. Van Dal, W.E. Wang, S. Hong, W. Vandervorst, K. De Meyer, S. Biesemans, P. Absil, N. Horiguchi and T. Hoffmann, *VLSI Tech. Digest*, 2011

14. M. Radosavljevic, G. Dewey, J.Fastenau, J. Kavalieros, R. Kotlyar, B. Chu-Kung, W. Liu, D.Lubyshev, M.Metz, K. Millard, N. Mukherjee, R. Pan, W. Pillarisetty, W. Rachmady, U. Shah and R. Chau, *IEDM Tech. Digest*, 2010

15. *International Technology Roadmap for Semiconductors*, 2011 Edition

16. A.D. Carter, J. Law, E. Lobisser, G. Burek, W. Mitchell, B. Thibeault, A. Gossard and M. Rodwell, *DRC Conference Proceedings*, pp. 19-20, 2011

17. A. Baraskar, V. Jain, M.A. Wistey, U. Singisetti, Y.U. Lee, B. Thibeault, A. Gossard and M. Rodwell, *IPRM Conference Proceedings*, July 2010

ECS Transactions, 45 (4) 129-136 (2012)
10.1149/1.3700461 ©The Electrochemical Society

Deterministic Assembly of $In_{0.53}Ga_{0.47}As$ p^+-i-n^+ Nanowire Junctions for Tunnel Transistors

M.-W. Kuo[a], J. Li[a], H. Liu[a], A. Vallett[a], D. K. Mohata[a], S. Datta[a], and T. S. Mayer[a,b]

[a] Department of Electrical Engineering, Penn State University,
University Park, Pennsylvania 16802, USA

[b] Department of Materials Science and Engineering, Penn State
University, University Park, Pennsylvania 16802, USA

Electric-field-assisted deterministic assembly is used to position arrays of p^+-i-n^+ $In_{0.53}Ga_{0.47}As$ nanowires on a Si substrate for device integration. Forces induced on the solution-suspended wires by a spatially varying, nonuniform electric field determines the position of each wire with respect to lithographic features on the substrate. The electrical properties of the sub-50 nm diameter $In_{0.53}Ga_{0.47}As$ junctions with a 100 nm thick unintentionally doped channel were characterized after adding source and drain contacts. The junctions showed clear rectification, with a forward bias ideality factors as low as 1.8, and reverse leakage current as small as 20 pA at a −1 V bias. This represents an important step towards demonstrating an $In_{0.53}Ga_{0.47}As$-based nanowire tunnel transistor.

Introduction

As metal oxide semiconductor field-effect transistor (MOSFET) power densities continue to rise at each technology node, tunnel field-effect transistors (TFETs) have emerged as an attractive low-power MOSFET replacement candidate.[1-7] In contrast to MOSFETs, TFETs have been shown to be able to break the 60 mV/dec subthreshold slope (SS) limit[3-4] because the gate controls tunneling through a barrier rather than emission over it, which filters out the high and low energy tails of the Fermi-Dirac distribution. Because nanowire devices have a circular geometry and a confined-volume body, a gate that wraps around the nanowire will provide excellent electrostatic control over the channel.[8-10] The improved channel control is expected to both reduce the S and improve the on-state current of nanowire TFETs.[7-8] Moreover, confinement effects such as the volume inversion of carriers[11] or the reduction of transverse momentum conservation requirements[12] may further enhance tunneling probabilities in nanowire systems.

Nanowires grown by the vapor-liquid-solid (VLS) method have the capability to reach diameters below 10 nm,[13] and also permit doped junctions and heterojunctions to be formed *in situ*, allowing for the synthesis of complex device structures. However, it has been difficult to achieve the extremely abrupt profile in the heavily doped source and

129

drain junction required for high performance devices using VLS growth. This paper describes the integration and characterization of $In_{0.53}Ga_{0.47}As$ p^+-i-n^+ nanowire junctions that are fabricated by high-aspect-ratio reactive ion etching (RIE) of device layers grown by molecular beam epitaxy (MBE). The dense arrays of sub-50 nm diameter vertically oriented nanowires are released from the InP growth substrate by selective etching, and are assembled into lateral device arrays on a Si substrate by electric-field assisted assembly. The source and drain metal contacts are deposited on the p^+ and n^+ segments of the wires, and the current-voltage (I-V) properties are measured under different ambient conditions. The etched junctions exhibit clear rectifying properties with a forward bias ideality factors as low as $n \sim 1.8$ and reverse leakage current as small as 20 pA at a −1V bias.

Experimental

The $In_{0.53}Ga_{0.47}As$ device structure shown in Figure 1a was grown by MBE on a lattice matched InP substrate using previously published conditions.[14] The symmetric structure consisted of a 250 nm thick n^+ drain layer doped with Si at density 10^{19} cm^{-3}, a 100 nm thick unintentionally doped channel layer, and a 250 nm thick p^+ source layer doped with C at a density of 10^{19} cm^{-3}. Following growth, vertically oriented nanowires were defined by lithographic patterning and high-aspect-ratio RIE. First, a 300 nm thick SiO_2 layer that served as a hard etch mask during nanowire RIE was deposited on the $In_{0.53}Ga_{0.47}As$ device structure by plasma enhanced chemical vapor deposition (PECVD). Interference lithography was then used to pattern a dense array of 80 nm diameter features in resist, and this pattern was transferred into the SiO_2 by RIE using CF_4 and CHF_3

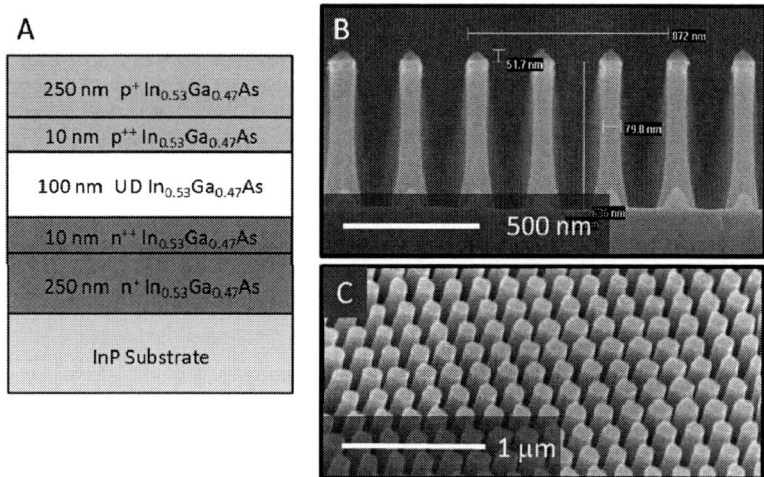

Figure 1. (a) $In_{0.53}Ga_{0.47}As$ p^+-i-n^+ device structure grown by MBE; (b) Cross-section FESEM image of 80 nm diameter nanowires fabricated by anisotropic RIE; (c) Top-view of dense array of nanowires patterned by interference lithography.

chemistry. Finally, the nanowires were etched by inductively coupled plasma (ICP) RIE using Cl_2, H_2, and Ar chemistry at substrate temperature of 200 °C.[15] Cross-sectional and top-view field emission scanning electron microscope (FESEM) images of the $In_{0.53}Ga_{0.47}As$ nanowires are shown in Figure 1b,c. High-aspect-ratio nanowires with diameters of 50 nm to 80 nm were obtained using this process.

To enable the integration of the $In_{0.53}Ga_{0.47}As$ nanowire junctions with Si CMOS, the etched wires were released from their InP growth substrate by selective wet etching and deterministically positioned into lateral device arrays by electric-field assisted assembly. As illustrated in Figure 2a, the InP substrate was removed using a highly selective isotropic dilute HCl (HCl:H_2O = 1.8:1) wet etch.[16] Following this selective etch, the substrate was rinsed in ethanol to remove the phosphoric acid by-product, and the loosely bound wires were gently sonicated into an isopropanol (IPA) solution. A FESEM image of the released nanowires is shown in Figure 2b.

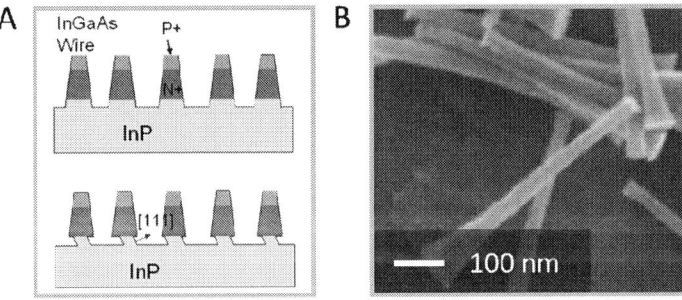

Figure 2. (a) Schematic diagram of the selective etching process using dilute HCl to undercut the InP substrate and release the $In_{0.53}Ga_{0.47}As$ nanowires; (b) FESEM image showing the released $In_{0.53}Ga_{0.47}As$ nanowires prior to deterministic assembly.

Individual wires were deterministically positioned in a regularly spaced array using electric-field assisted assembly.[17-19] In this technique, the solution-suspended polarizable nanowires are aligned and attracted to predefined locations on the Si substrate with submicron registration accuracy to existing lithographic features on the substrate. This is due to the dielectrophoretic (DEP) force that is induced on the polarized wires by a non-uniform alternating electric field supplied to the IPA solution via dielectric-coated metal electrodes on the substrate. This long-range force causes the wires to align tangent to the electric field lines and to move in the direction of the highest electric field gradient.[20] Localized regions of high electric field strength are obtained by lithographically patterning a regular array of depressions in a sacrificial PMGI dielectric layer that coats the interdigitated electrode structure (see Figure 3a). The three-dimensional electric field simulation results shown in Figure 3b,c demonstrate that the highest electric field strength is found within the lithographic depressions, and that the field gradient extends into the solution.

This causes an individual nanowire to preferentially assemble within each depression. Once assembled, the short-range electric-field forces between the assembled wires and the underlying electrode structure result in end-to-end alignment of the nanowires with respect to the electrode edge. This provides the registration accuracy needed to overlay the source and drain contacts to the 250 nm long source and drain segments without shorting out the channel region.

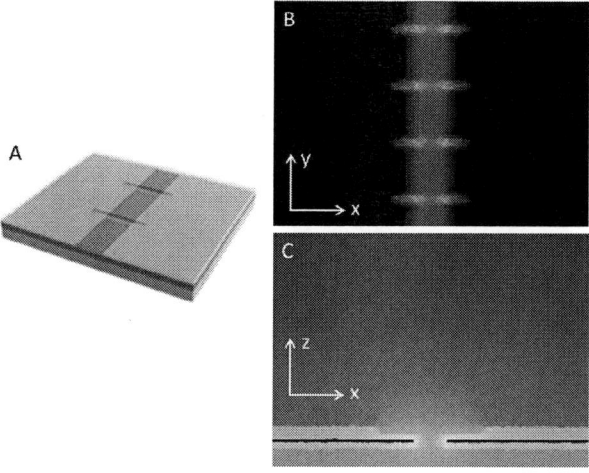

Figure 3. (a) Schematic illustration of a pair of interdigitated assembly electrodes with lithographic depressions; (b) Top-view (x-y plane) of electric-field intensity. The highest intensity (light) is within the depressions. (c) Cross-sectional view (x-z plane) of the electric field intensity. The field gradient extends into the solution.

The field-assisted assembly was conducted by applying a 6 $V_{peak-to-peak}$, 10 kHz sinusoidal voltage across the interdigitated electrodes, while flowing the nanowire solution across the substrate surface. Figure 4a shows a FESEM image of the deterministically assembled $In_{0.53}Ga_{0.47}As$ nanowire junctions. Importantly, the spacing between the assembled wires in the array is well controlled, and the ends of the adjacent wires are aligned. In addition, the junctions preferentially orient such that the p^+ segments face the same direction.[21] Following assembly, the source and drain contacts windows were defined by electron-beam lithography in ZEP 520 resist. The native oxide on the surface of the wire in the exposed windows was removed in highly diluted HCl (HCl:H_2O = 1:50), and the 150 nm Ti/20 nm Pd contact metal was subsequently deposited by electron-beam evaporation. The device fabrication was completed by dissolving the remaining resist in n,n-Dimethylacetamide, which resulted in an array of electrically isolated nanowire junction devices. A high-magnification FESEM image taken before and after adding the source and drain contacts to a single 70 nm diameter $In_{0.53}Ga_{0.47}As$ nanowire junction is shown in Figure 4b,c. Each nanowire segment is labeled in Figure 4b, showing that the

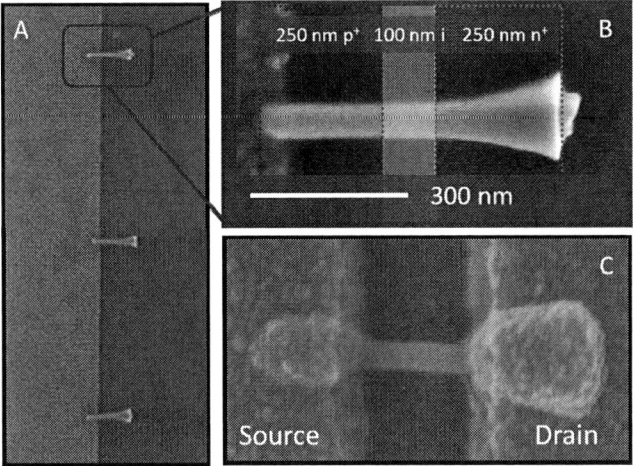

Figure 4. (a) Uniformly spaced array of individual $In_{0.53}Ga_{0.47}As$ nanowires following assembly. An $In_{0.53}Ga_{0.47}As$ nanowire (b) before and (c) after patterning and depositing the source and drain metal contacts.

junction is centered between the two metal contacts. The I-V properties of these junctions were characterized following a 20 min. contact anneal at 300°C in 5% H_2/N_2.

Results and Discussion

The room-temperature I-V properties of individual unpassivated 50 nm to 80 nm diameter $In_{0.53}Ga_{0.47}As$ nanowire junctions were measured using a Lakeshore Cryogenic probe system. The devices were placed in the enclosed chamber of the cryo-prober, and measurements were taken on the same device at each step in the following sequence: (1) purging with dry N_2 for 12 hours, (2) evacuating to 10^{-3} Torr, (3) evacuating to 10^{-6} Torr for 2 hours, (4) backfilling with room air, and (5) purging with dry N_2. Figure 5 shows a plot of the current density (J)-V for a typical nanowire junction device measured under each condition. These results demonstrate that the reverse bias current density is the lowest (2×10^{-1} A/cm^2) after the device is purged in N_2 for 12 hours, and increases by more than a factor of ten when the chamber is evacuated to 10^{-7} Torr. The current density remains high when the device is exposed to air, and returns to the lowest value after subsequent purging in N_2. The forward bias ideality factor is $n \sim 1.8$ in N_2, which is consistent with SRH-dominated recombination in the depletion region. A significant increase in the leakage current at low values of forward bias is observed for all other conditions.

Large-area junctions (11×20 μm^2) fabricated using a similar device structure have demonstrated a similar reverse leakage current density but less environmental dependence, and were dominated by recombination/generation in the bulk.[14] In contrast, these results indicate that current in the $In_{0.53}Ga_{0.47}As$ p$^+$-i-n$^+$ nanowire junctions is dominated

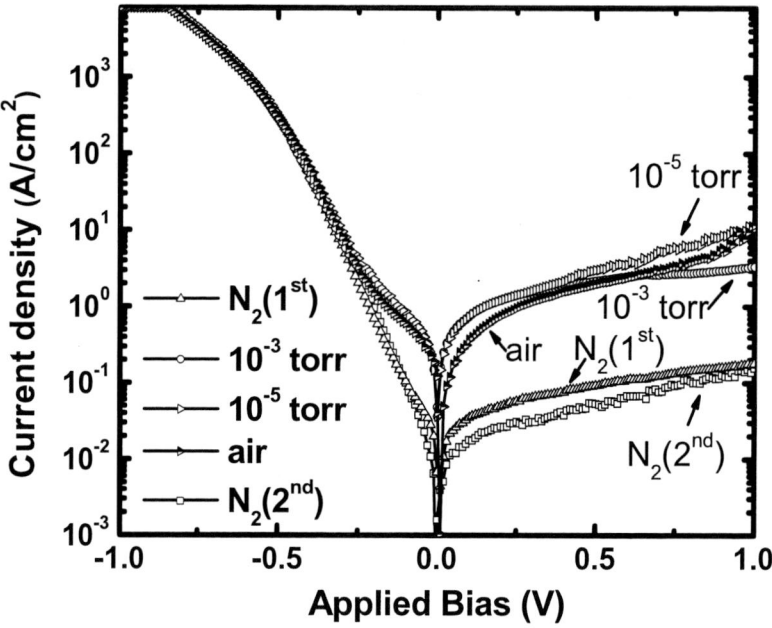

Figure 5. Current density as a function of applied bias measured in a cryogenic prober on the same $In_{0.53}Ga_{0.47}As$ p^+-i-n^+ nanowire junction at different ambient conditions.

by the surface rather than the bulk. This difference can be attributed to two primary factors. First, the large-area junctions were isolated using a wet etch rather than the dry etch used here. It is well known that surface damage is introduced during RIE of III-V materials, and that this damage increases in the surface state density.[22] Second, the perimeter-to-area ratio of the nanowire devices is more than 200 times larger than the wet etched junctions, which results in an increased sensitivity to the surface. The reproducible changes in I-V characteristics further suggests that the ambient measurement conditions modulate the surface band bending, and hence the recombination/generation current. Despite the increased leakage current as compared to the large-area devices, these abruptly doped $In_{0.53}Ga_{0.47}As$ p^+-i-n^+ nanowire junctions show clear rectification and low reverse leakage current. Ongoing work is underway to study the dry etched surface using X-ray photoelectron spectroscopy, and to develop effective methods to passivate this surface with high-κ gate dielectrics for tunnel transistor integration.

Conclusions

This paper discussed the integration of p^+-i-n^+ $In_{0.53}Ga_{0.47}As$ nanowire junction arrays on a Si substrate. The 50 nm to 80 nm diameter wires were fabricated by high-aspect-ratio dry etching of a MBE-grown device structure with abruptly doped p^+-i and i-n^+ junctions. The etched wires were released from the growth substrate, and were deterministically positioned with submicron registration accuracy relative to lithographic features on the Si substrate using electric-field assisted assembly. The spacing between the wires in the array was well controlled, the ends of adjacent wires were aligned, and the p^+ segments were preferentially oriented in the same direction. Nanowire junction devices were fabricated from the assembled wires by lithographically defining source and drain metal contacts on the p^+ and n^+ wire segments. The room-temperature I-V characteristics of the unpassivated p^+-i-n^+ $In_{0.53}Ga_{0.47}As$ junctions were strongly rectifying, but depended on the ambient measurement conditions. The electrical measurement results suggest the increased leakage current as compared to large area wet etched devices is due to surface-dominated recombination/generation current, which may be due to dry-etch induced surface damage and the large perimeter-to area ratio of the nanowire devices. Future work will investigate strategies to remove the dry etch damage and to passivate the nanowire surface with high-κ gate dielectrics for tunnel transistor integration.

Acknowledgments

This work was supported by the Midwest Institute for Nanoelectronics Discovery (MIND) and the National Science Foundation. The authors acknowledge the growth of the epitaxial device layers by IQE, Inc., and the high-aspect-ratio nanowire RIE by N. Cao at the UCSB National Nanotechnology Infrastructure Network (NNIN) Site. The facilities at the Penn State Site of the NNIN were used to fabricate the nanowire devices.

References

[1] K. K. Bhuwalka, S. Sedlmaier, A. K. Ludsteck, C. Tolksdorf, and J. Schulze, *IEEE Transactions on Electron Devices*, **51**(2) 279 (2004).

[2] J. Appenzeller, Y. M. Lin, J. Knoch, and P. Avouris, *Physical Review Letters*, **93**(19) 196805 (2004).

[3] W. Y. Choi, B.-G. Park, J. D. Lee, and T.-J. K. Liu, *Electron Device Letters, IEEE*, **28**(8) 743 (2007).

[4] S. O. Koswatta, M. S. Lundstrom, and D. E. Nikonov, *IEEE Transactions on Electron Devices*, **56**(3) 456 (2009).

[5] Y. Khatami and K. Banerjee, *IEEE Trans. Electron Devices*, **56**(11) 2752 (2009).

[6] J. Appenzeller, J. Knoch, M. T. Bjork, H. Riel, H. Schmid, and W. Riess, *IEEE Trans. Electron Devices*, **55**(11) 2827 (2008).

[7] C. Thelander, et al., *Mater. Today*, **9**(10) 28 (2006).

[8] J. Guo, J. Wang, E. Polizzi, S. Datta, and M. Lundstrom, *IEEE Transactions on Nanotechnology*, **2**(4) 329 (2003).

[9] A. S. Verhulst, B. Soree, D. Leonelli, W. G. Vandenberghe, and G. Groeseneken, *J. Appl. Phys.*, **107**(2) 024518 (2010).

[10] J. P. Colinge, M. H. Gao, A. Romano-Rodriguez, H. Maes, and C. Claeys, *IEEE International Electron Devices Meeting (IEDM)*, 595 (1990).

[11] J. Knoch, S. Mantl, and J. Appenzeller, *Solid-State Electronics*, **51**(4) 572 (2007).

[12] M. T. Bjork, J. Knoch, H. Schmid, H. Riel, and W. Riess, *Appl. Phys. Lett.*, **92**(19) 193504 (2008).

[13] C. Yang, Z. Zhong, and C. M. Lieber, *Science*, **310**(5752) 1304 (2005).

[14] S. Mookerjea, et al., *IEEE International Electron Devices Meeting (IEDM)*, 316 (2009).

[15] J. S. Parker, E. J. Norberg, H. Yung-Jr, K. Byungchae, R. S. Guzzon, and L. A. Coldren, *IEEE Photonics Technology Letters*, **23**(9) 573 (2011).

[16] S. Arscott, P. Mounaix, and D. Lippens, *J. Vacuum Science & Technology B*, **18**(1) 150 (2000).

[17] P. A. Smith, C. D. Nordquist, T. N. Jackson, T. S. Mayer, B. R. Martin, J. Mbindyo, and T. E. Mallouk, *Appl. Phys. Lett.*, **77**(9) 1399 (2000).

[18] T. J. Morrow, M. Li, J. Kim, T. S. Mayer, and C. D. Keating, *Science*, **323**(5912) 352 (2009).

[19] M. Li, R. B. Bhiladvala, T. J. Morrow, J. A. Sioss, K.-K. Lew, J. M. Redwing, C. D. Keating, and T. S. Mayer, *Nat. Nano.*, **3**(2) 88 (2008).

[20] B. D. Smith, T. S. Mayer, and C. D. Keating, *Annual Reviews in Physical Chemistry*, in press.

[21] C. H. Lee, D. R. Kim, and X. Zheng, *Nano Letters*, **10**(12) 5116 (2010).

[22] E. L. Hu and C.-H. Chen, *Microelectronic Engineering*, **35**(1–4) 23 (1997).

Desorption of Ge species during thermal oxidation of Ge and annealing of HfO$_2$/GeO$_2$ stacks

C. Radtke[a], G.K. Rolim[a], S.R.M. da Silva[b], G.V. Soares[b], C. Krug[b,c], and I.J.R. Baumvol[b,d]

[a] Instituto de Química and [b] Instituto de Física, UFRGS, Porto Alegre, 91509-900, Brazil
[c] CEITEC S.A., Porto Alegre, 91550-000, Brazil
[d] Universidade de Caxias do Sul, Caxias do Sul, 95070-560, Brazil

Thermal stability of HfO$_2$/GeO$_2$ stacks was investigated. These structures were stable on Si up to 600 °C following annealing in N$_2$. Samples prepared on Ge yielded direct evidence of migration and loss of both Ge and O. This contrasting behavior is a result of GeO formation at the GeO$_2$/Ge interface. Desorption of Ge was also observed during thermal oxidation of this semiconductor. The amounts of desorbed Ge were related to the oxygen pressure. Oxygen isotopic tracing results evidenced a strong interaction of oxygen from the gas phase with the already formed GeO$_2$ layer.

Introduction

Germanium (Ge) is being considered a promising candidate as a replacement for silicon (Si) as the channel material of metal-oxide-semiconductor field-effect transistors (MOSFETs). This semiconductor material has the highest hole mobility, which is about twofold higher than the best III-V-based p-type material. Furthermore, it has an adequate bandgap for future non-silicon CMOS technologies (1-3). However, finding an efficient passivation scheme for Ge is still a challenge. A considerable percentage of the intrinsically high carrier mobility of Ge can be only attained in devices if an adequate gate insulator material is formed on its surface. This layer must form an interface with Ge with a sufficiently low density of active interface defects (4). The formed dielectric/Ge structure must also withstand the transistor fabrication process which involves anneals at temperatures above 400 °C (5).

Unlike silicon oxide (SiO$_2$), Ge native oxide is unstable and water soluble. These characteristics constitute serious drawbacks in terms of device processing and electrical properties of the formed MOS structure. GeO desorption and subsequent incorporation in deposited dielectric layers can degrade the electrical properties of dielectric/Ge stacks (5). Thus, the interaction of these layers with Ge is a key issue in the fabrication of Ge MOSFETs. Kita et al. (6) showed that below 700 °C GeO desorption is not a result of GeO$_2$ decomposition itself but rather from its reaction with the underlying Ge substrate. Using isotopic tracing techniques, Wang et al. (7) proposed an oxygen vacancy diffusion model to explain GeO desorption: oxygen vacancies generated at the GeO$_2$/Ge interface diffuse into the GeO$_2$ surface region resulting in GeO desorption.

Due to current device scaling, the incorporation of high-k gate-dielectric materials such as HfO_2 is mandatory if Ge MOSFETs are to become commercially relevant. However, Ge substrate oxidation can occur during deposition of the dielectric material and/or device processing (8-10). Atomic layer deposition (ALD) of HfO_2 on a Ge surface after wet chemical treatment resulted in quasiepitaxial growth whereas a nitrided Ge surface resulted in an amorphous layer (11). Ge incorporation inside the dielectric layer was observed with time-of-flight secondary ion mass spectrometry (TOF-SIMS) (12) and was related to an increase in interface trap density. These and other results evidence deleterious effects of the direct deposition of HfO_2 on Ge pointing to the need of interlayers between these materials (13). It remains to be shown if an intentionally grown GeO_2 layer (as opposed to one generated as by-product during processing) can fulfill this role.

In view of this scenario, we compared HfO_2/GeO_2 stacks on Ge and Si with respect to thermally driven atomic transport. The combined use of ion beam analysis and photoelectron spectroscopy indicates that Ge from the substrate drives instabilities observed at the HfO_2/GeO_2 interface. In order to have a deeper insight into the interaction of GeO_2 with Ge, we also investigated desorption occurring during thermal oxidation of Ge. Our results support the proposal of oxygen vacancy diffusion during the process.

Experimental

p-type substrates were epiready Ge (100) wafers doped with Ga (Umicore), with a resistivity of 0.24-0.47 ohm.cm. They were first cleaned in an ultrasonic acetone bath, and then etched in a 40% HF aqueous solution for 1 min. Si (100) samples were cleaned in a mixture of H_2SO_4 and H_2O_2 followed by the same procedure used for Ge substrates. After rinsing the samples in deionized water, they were immediately transferred to load lock chambers. HfO_2/GeO_2 stacks were prepared on both Si and Ge substrates. Si samples were loaded in a remote plasma enhanced chemical vapor deposition (RPECVD) reactor, where a 5 nm layer of GeO_2 was deposited followed by a 3 nm layer of HfO_2. Ge substrates were loaded in a resistively heated quartz tube furnace that was pumped down to 2×10^{-7} mbar and then pressurized with 200 mbar of O_2 enriched to 97% in the ^{18}O rare isotope (termed $^{18}O_2$). Ge thermal oxidation was performed at 450 °C for 2h, leading to a 5 nm $Ge^{18}O_2$ layer. Following this step, a 3 nm HfO_2 layer was deposited by RPECVD. Post-deposition annealings (PDAs) of the resulting $HfO_2/Ge^{18}O_2/Ge$ and $HfO_2/Ge^{18}O_2/Si$ structures were performed in 1 atm of N_2 at 500 or 600 °C for 1 to 4 h.

Ge desorption was also investigated during thermal oxidation of bare Ge substrates. Sequential oxidations using natural oxygen (termed O_2) and $^{18}O_2$ were performed aiming at identifying the interaction of oxygen from the gas phase with the GeO_2/Ge structure. Ge desorption during GeO_2 thermal growth was tracked by placing a thermally oxidized Si sample above the Ge substrate (14). Depth distributions of ^{18}O were obtained by nuclear reaction profiling (NRP) using the resonance at 151 keV in the cross section curve of the $^{18}O(p,\alpha)^{15}N$ nuclear reaction (15). X-ray photoelectron spectroscopy (XPS) was performed using Al $K\alpha$ radiation. Ge and Hf amounts were determined by Rutherford backscattering spectrometry (RBS) using He^+ ions of 1 MeV.

Results and Discussion

Figure 1 shows Hf 4f XPS spectra for samples prepared on Ge and on Si, both as-deposited and annealed at 600 °C for 1h. The spectral component at a binding energy (BE) around 18.2 eV (Hf $4f_{7/2}$) can be assigned to Hf-O bonding in stoichiometric HfO_2 (16), indicating that HfO_2 and the underlying GeO_2 do not react during RPECVD deposition at 300 °C. After N_2 annealing, only Hf-O bonding was observed in all samples annealed at 500 °C and in the Si sample annealed at 600 °C. Ge samples annealed at 600 °C (irrespective of time) showed an additional spectral component assigned to Hf-O-Ge bonding (17). This result indicates that the substrate onto which the HfO_2/GeO_2 stacks are deposited plays a major role regarding thermal stability. The process that triggers chemical interaction between HfO_2 and GeO_2 on Ge above 500 °C is not observed on Si.

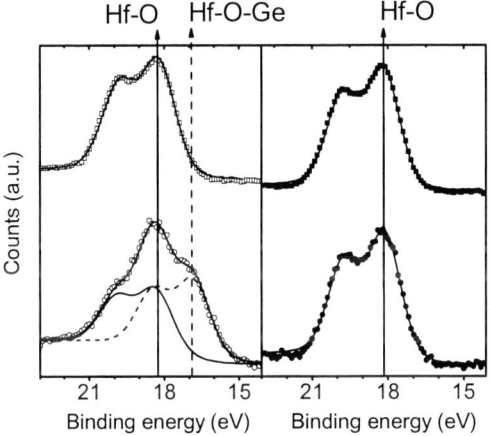

Figure 1. Hf 4f XPS spectra of the HfO_2/GeO_2 stacks on Ge (left-hand side and open symbols) and on Si (right-hand side and full symbols). Squares correspond to as-deposited samples and circles to those annealed at 600 °C for 1h. The binding energy of spectral components assigned to Hf-O and Hf-O-Ge is indicated. a.u. stands for arbitrary units.

Fig. 2 shows ^{18}O concentration profiles obtained by NRP applied to $HfO_2/Ge^{18}O_2/Ge$ structures submitted to PDAs in N_2 at 500 °C for 1 and 4 h. The growth of a GeO_2 layer enriched in the ^{18}O isotope enables the use of NRP to profile oxygen with subnanometer depth resolution and also to distinguish oxygen from the GeO_2 layer from that of HfO_2. The as-deposited sample presents a box-like ^{18}O profile compatible with 5 nm of $Ge^{18}O_2$ under 3 nm of HfO_2. Two different effects are observed following annealings at 500 °C: (i) reduction of the ^{18}O concentration in both samples by similar amounts and (ii) incorporation of ^{18}O to the original HfO_2

overlayer. Regarding (i), previous works (7,18) showed that annealing the GeO_2/Ge structure may lead to a solid state reaction producing GeO. The latter, volatile compound can diffuse to the ambient. Thus, we attribute the decrease in ^{18}O concentration to volatilization of $Ge^{18}O$. Concerning ^{18}O incorporation in HfO_2, we suggest that this observation is due to trapping of $Ge^{18}O$ during transport to the sample surface. Thus, (i) and (ii) would be directly related.

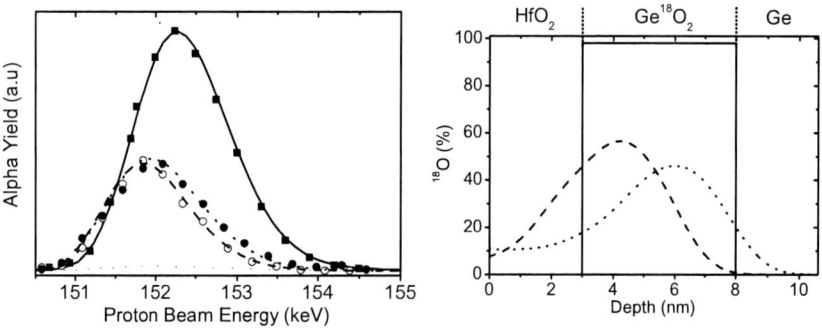

Figure 2. (Left-hand side) Excitation curves (symbols) of the $^{18}O(p,\alpha)^{15}N$ nuclear reaction around the resonance energy $E_r = 151$ keV and the corresponding simulation (lines) for the as-deposited sample (black full squares and solid line) and for samples annealed at 500 °C for 1 h (blue full circles and dotted line) or 4 h (red open circles and dashed line). (Right-hand side) ^{18}O concentration profiles assumed in the simulations. a.u. stands for arbitrary units.

In order to investigate GeO_2 removal from the HfO_2/GeO_2 stacks following PDAs in N_2, we performed Rutherford backscattering spectrometry in channeling geometry (c-RBS). By this way, we were able to distinguish between Ge from the crystalline substrate and Ge from the GeO_2 layer and/or incorporated in HfO_2. Fig. 3 shows the reminiscent amounts of Ge inside the dielectric layer following PDAs. As before, no instability was observed in Si samples. Regarding the Ge substrate, annealing at 500 °C (Fig. 3(a)) led to the loss of Ge in the initial GeO_2 film, which again supports volatilization of GeO. PDAs at 600 °C (Fig. 3(b)) originally produced a puzzling result: a significant increase in the amount of Ge in the dielectric stack. Previous works (12,19) indicate the possibility of surface deposition of GeO generated in the annealing chamber due to oxidation of the backside of the Ge substrate. In this work, such oxidation would be due to residual oxygen in the furnace. To confirm this hypothesis we performed the same processing at 600 °C using Ge samples whose back side had been covered with 300 nm of Si. As shown in Fig. 3(b), Ge loss became clearly detectable, pointing to accelerated evolution of GeO at 600 °C.

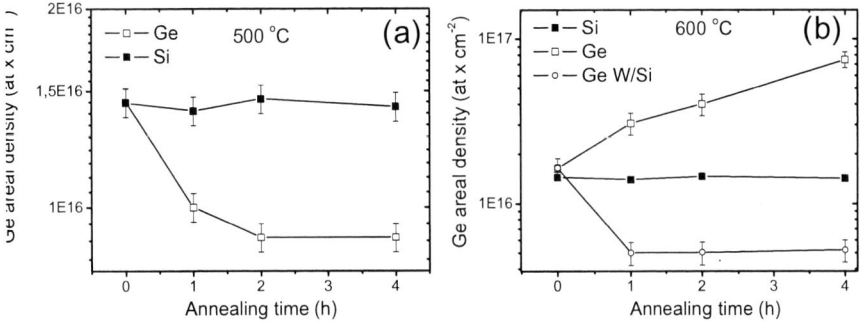

Figure 3. Ge areal densities obtained from c-RBS as a function of annealing time of samples deposited on Si (full squares), Ge (open squares), and Ge with a Si back layer (open circles). PDAs were performed at (a) 500 °C and (b) 600 °C.

Desorption of Ge related species was also investigated during the thermal growth of GeO_2 layers on Ge. The absence of the HfO_2 capping layer enabled the investigation of the influence of oxygen pressure on the stability of GeO_2. Understanding this process is a key issue in order to control the properties of the formed GeO_2 layer which could be useful as an interlayer between Ge and high-k dielectrics like HfO_2. A thermally oxidized Si sample was placed above the Ge substrate in this experiment. Fig. 4 shows the RBS spectrum of this Si sample for a specific oxidation condition. A clear signal of Ge is observed, evidencing that Ge related species desorb during thermal oxidation of Ge. The amount of Ge adsorbed on the Si sample was 66×10^{15} Ge·cm^{-2}. This amount varies as different oxidation parameters are employed. Lower oxygen pressures (100 and 200 mbar) result in higher Ge amounts (562×10^{15} Ge·cm^{-2} and 471×10^{15} Ge·cm^{-2}, respectively).

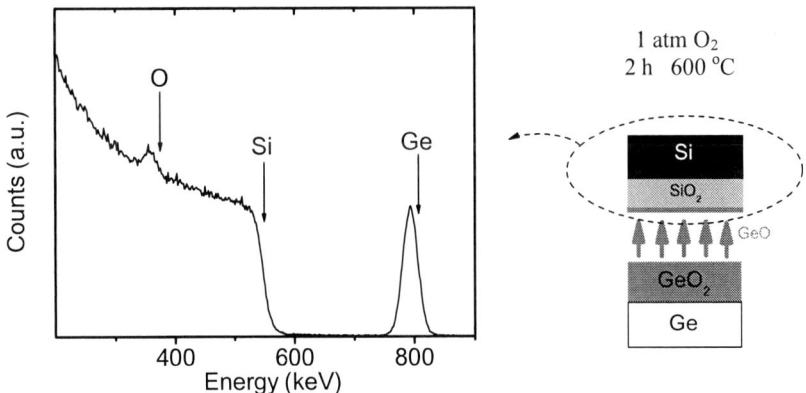

Figure 4. RBS spectrum from a SiO_2/Si sample placed above a Ge substrate during thermal oxidation. Oxidation parameters are listed beside the spectrum. A sketch of the experimental setup is also presented. Arrows in the graph indicate the signals from Ge, Si, and O. a.u. stands for arbitrary units.

Oxygen transport and incorporation occurring during thermal oxidation of Ge were investigated with O isotopic labeling in conjunction with subnanometer ^{18}O depth profiling. Fig. 5 shows nuclear reaction data regarding ^{18}O in Ge substrates following thermal oxidation. Alpha particle yield in the figure is proportional to ^{18}O concentration; depth in the sample scales with proton energy. The curves correspond to the actual concentration versus depth information convoluted with instrumental and proton energy loss functions. Two samples were prepared. Sample A was a Ge substrate thermally oxidized in ^{18}O$_2$. Sample B was another Ge substrate which underwent two sequential oxidations: the first one in natural oxygen, followed by a second one in ^{18}O$_2$. By this way, we were able to follow the interaction of O from the gas phase with the GeO$_2$/Ge structure. Results of sample A evidence the formation of a Ge^{18}O$_2$ layer, as expected. The alpha yield in the plateau region of the respective excitation curve corresponds to the oxygen concentration in stoichiometric GeO$_2$. In the case of sample B, ^{18}O reaches deeper regions as a result of GeO$_2$ growth during the second oxidation step in ^{18}O$_2$. ^{18}O incorporation is observed throughout the GeO$_2$ layer evidencing that oxygen from the gas phase incorporates in the previously formed GeO$_2$ layer besides reacting with the Ge substrate forming new oxide. This result can be explained on the basis of the model proposed by Wang et al. (7) of oxygen vacancies diffusion from the GeO$_2$/Ge interface towards the GeO$_2$ surface region. During thermal oxidation, O mobility in GeO$_2$ is promoted by these vacancies, resulting in the observed ^{18}O distribution. This mechanism is different from that observed for the SiO$_2$/Si system, where oxygen from the gas phase diffuses through the growing oxide film (without interacting with it) and reacts with the semiconductor substrate, forming SiO$_2$ (20).

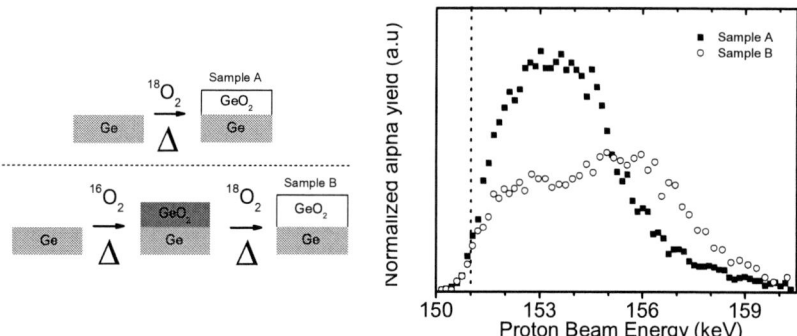

Figure 5. Experimental excitation curves of the ^{18}O(p,α)^{15}N nuclear reaction. Sample A was prepared by a single oxidation step in 1 atm of ^{18}O$_2$ for 15 min at 600 °C. Sample B was prepared in two oxidation steps: the first one was performed in natural O$_2$ with the same conditions used in sample A, followed by a second step performed in 1 atm of ^{18}O$_2$ at the same conditions for 30 min. A sketch of sample preparation is shown in the left side of the figure. The vertical line at 151 keV indicates the energy position corresponding to ^{18}O at the sample surface. a.u. stands for arbitrary units.

Conclusions

The results evidence that the HfO_2/GeO_2 stack deposited on Si is stable up to 600 °C following annealing in N_2. Samples prepared on Ge yielded direct evidence of migration and loss of both Ge and O. This contrasting behavior is a result of GeO formation at the GeO_2/Ge interface. Despite the transport of Ge into HfO_2 already at 500 °C, effective formation of germanates occurred only at 600°C as evidenced by XPS results. Desorption of Ge was also observed during thermal oxidation of this semiconductor. The amounts of desorbed Ge are related to oxygen pressure. Oxygen isotopic tracing results evidenced a strong interaction of oxygen from the gas with GeO_2 on the Ge substrate. This result is in sharp contrast to similar experiments performed with Si samples, evidencing completely different mechanisms involved in the thermal growth of oxide layers on these semiconductor materials.

Acknowledgments

Centro de Microscopia Eletrônica CME-UFRGS, INCT Namitec, INCT INES, MCT/CNPq, CAPES, and FAPERGS.

References

1. R. Pillarisetty, Nature **479**, 324 (2011).
2. M. Heyns and W. Tsai, MRS Bull. **34**, 485 (2009).
3. A. Lubow, S. Ismail-Beigi, and T.P.Ma, Appl. Phys. Lett. **96**, 122105 (2010).
4. R.M. Wallace, Paul C. McIntyre, J. Kim, and Y. Nishi, MRS Bull. **34**, 493 (2009).
5. Y. Kamata, Mater. Today **11**, 30 (2008).
6. K. Kita, S. Suzuki, H. Nomura, T. Takahashi, T. Nishimura, and A. Toriumi, Jpn. J. Appl. Phys. **47**, 2349 (2008).
7. S.K.Wang, K.Kita, T. Nishimura, K. Nagashio, and A.Toriumi, Jpn. J. Appl. Phys. **50**, 04DA01 (2011).
8. C. Radtke, N.M. Bom, G.V. Soares, C. Krug, and I.J.R. Baumvol, ECS Trans. 41 (3), 21 (2011).
9. C. Radtke, C. Krug, G.V. Soares, I.J.R. Baumvol, J.M.J. Lopes, E. Durgun-Ozben, A. Nichau, J. Schubert, and S. Mantl, Electrochem. Solid-State Lett. **13**, G37 (2010).
10. C. Henkel, O. Bethge, S. Abermann, S. Puchner, H. Hutter, and E. Bertagnolli, Appl. Phys. Lett. **91**, 82904 (2007).
11. E.P.Gusev, H. Shang, M. Copel, M. Gribelyuk, C. D'Emic, P. Kozlowski, and T. Zabel, Appl. Phys. Lett. **85**, 2334 (2004).
12. Q. Zhang, N. Wu, D.M.Y. Lai, Y. Nikolai, L.K.Bera, and C. Zhu, J. Electrochem. Soc. **153**, G207 (2006).
13. A.Delabie, F. Bellenger, M. Houssa, T. Conard, S.V. Elshocht, M. Caymax, M. Heyns, and M. Meuris, Appl. Phys. Lett. **91**, 82904 (2007).
14. Y. Oniki, H. Koumo, Y. Iwazaki, and T. Ueno, J. Appl. Phys. **107**, 124113 (2010).
15. C. Driemeier, L. Miotti, R.P. Pezzi, K.P. Bastos, and I.J.R. Baumvol, Nucl. Instrum. Methods Phys. Res. B **249**, 278 (2006).
16. C. Morant, L. Galán, and J.M. Sanz, Surf. Interface Anal. **16**, 304 (1990).

17. S.Van Elshocht, M. Caymax, T. Conard, S. De Gent, I. Hoflijk, M. Houssa, B. De Jaeger, J. Van Sttebergen, M. Heyns, and M. Meuris, Appl. Phys. Lett. **88**, 141904 (2006).
18. N.Lu, W. Bai, A. Ramirez, C. Mouli, A. Ritenour, M.L.Lee, D. Antoniadis, and D.L. Kwong, Appl. Phys. Lett. **87**, 051922 (2005).
19. C.-C. Cheng, C.-H. Chien, G.-L. Luo, C.-H. Yang, M.-L. Kuo, J.-H. Lin, and C.-Y. Chang, A, Appl. Phys. Lett. **90**, 012905 (2007).
20. I.J.R. Baumvol, Surf. Sci. Rep. **36**, 1 (1999).

CHAPTER 5

III-V INTEGRATION

ECS Transactions, 45 (4) 147-149 (2012)
10.1149/1.3700464 ©The Electrochemical Society

New Method to Produce High-Quality Epitaxial Ge on Si Using SiO₂-Lined Etch Pits and Epitaxial Lateral Overgrowth for III-V Integration

Darin Leonhardt[1] and Sang M. Han[2]

[1] Sandia National Laboratories, Albuquerque, NM 87185
[2] Department of Chemical & Nuclear Engineering
University of New Mexico, Albuquerque, NM 87131

We have developed a unique method of reducing the threading dislocation density in Ge epitaxially grown on Si. A heteroepitaxial layer of Ge is initially grown on Si substrate in which the threading dislocation density is 2.6×10^8 cm^{-2}. The Ge film is etched to produce pits on the surface that correspond to the location of threading dislocations. The Ge surface with etch pits is processed further to produce a 15 nm thick layer of SiO₂ within the etch pits, but not on the top planar Ge film surface. Further lateral and selective epitaxial Ge growth results in a Ge film that forms and coalesces over the SiO₂ lined etch pits. The SiO₂ blocks the threading dislocations from propagating into the subsequent Ge epilayer. Etching the subsequent Ge epilayer reveals a density of 1.7×10^6 cm^{-2} threading dislocations. This corresponds to a decrease in the overall defect density by a factor of 31 compared to the initial Ge on Si layer. The threading dislocations likely result from shallow etch pits in which SiO₂ is inadvertently removed from the pit during one of the processing steps. The dislocations in pits without SiO₂ continue to propagate into the subsequent Ge epilayer. Reducing the dislocation density in the initial Ge layer should result in fewer defects in the subsequent Ge layer, making the layer suitable for electronic device fabrication.

Introduction:

Integrating a high-quality layer of epitaxial Ge on Si has been a longstanding engineering challenge, despite its technological importance. The applications of Ge-on-Si include high-mobility field-effect transistors[1,2], 'virtual substrates' for III-V multijunction solar cells[3], and optical interconnects monolithically integrated with Si-based circuitry[4]. The primary difficulties in achieving Ge films of sufficient quality stem from the lattice mismatch that leads to a large density ($> 10^9$ cm^{-2}) of threading dislocations (TDs) and the thermal expansion coefficient mismatch between Ge and Si that leads to microcracks or delamination of Ge film upon cooling from growth to room temperature.

Experimental Results:

Herein, we present a new method to reduce the TD density, using a minimal number of standard microfabrication steps. Figure 1 shows our process flow. The method begins with growing a 500-nm-thick epitaxial Ge layer on Si. A post-growth anneal step leads to a TD density of approximately 5×10^7 cm^{-2}, as revealed by plan-view transmission electron microscopy (TEM) and etch pit density (EPD) measurements. The close agreement between EPD measurements and TEM shows that the EPD measurements reliably decorate all TDs. Etch pits are created around the dislocation cores in the Ge film. A 15-nm-thick layer of SiO_2 is subsequently deposited on the etch-pit-decorated Ge film. A thin layer of polymethyl methacrylate (PMMA) is then spin-coated onto the sample, which fills the etch pits and planarizes the Ge surface. Next, a reactive ion etching step is used to remove the polymer and SiO_2 from the planar regions of the sample surface surrounding the etch pits. An O_2 plasma is then used to selectively remove the remaining polymer, so that SiO_2 remains only within the etch pits. Lastly, a second layer of Ge is selectively grown on the exposed Ge surface and laterally over the SiO_2-lined etch pits until a fully coalesced Ge film is created. A final polishing step produces an atomically flat continuous Ge film.

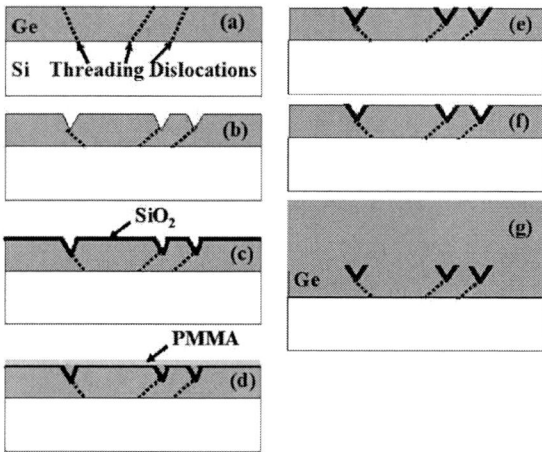

Figure 1: Process flow diagram for SiO_2-lined etch pit technique to block threading dislocations in epitaxially grown Ge on Si.

Analysis:

A cross-sectional TEM image in Figure 2 shows the resulting architecture. At high growth temperatures, shown by Figure 2(a), the Ge epilayer can be selectively grown over the SiO_2, and the resulting coalescence of merging Ge over the SiO_2 can lead to relatively defect-free Ge. In comparison, at low growth temperatures, the Ge coalescence leads to random nucleation of Ge on top of SiO_2 and formation of a polycrystalline film as shown by Figure 2(b). Ensuing EPD measurements reveal that the density of twin defects and TDs in the upper Ge layer is approximately 1.7×10^6 cm^{-2}, such that the overall defect density is reduced by a factor greater than 30 compared to that in the initial Ge layer. Both theoretical and experimental evidence suggest that the defect density in GaAs films on Ge/Si must be less than 2×10^6 cm^{-2} to have a minority carrier lifetime comparable to GaAs films grown on Ge and GaAs substrates.[5] Therefore, our new method of using SiO_2-lined etch pits to block the propagation of TDs in Ge may finally lead to device quality III-V materials integrated on Si substrates.

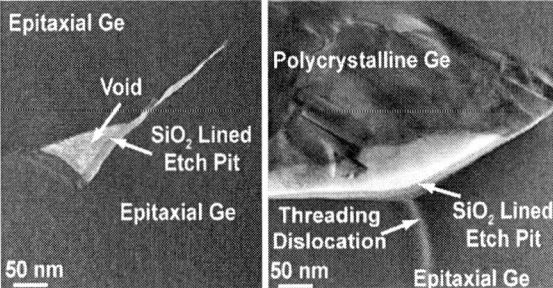

Figure 2: Cross-sectional TEM image capturing SiO_2-lined etch pits and TD termination a) at high growth temperatures and b) at low growth temperatures.

In summary, an etch-pit plugging method is considered to reduce the threading dislocation density caused by lattice mismatch between Ge and Si. This corresponds to a decrease in the overall defect density by a factor of 31 compared to the initial Ge on Si layer.

Acknowledgments

The above material is based upon work supported by, or in part by, the National Science Foundation (DMR-0907112 and CMMI-1068970). The authors also acknowledge the use of TEM facility at the University of New Mexico.

References

[1] D. P. Brunco, B. De Jaeger, G. Eneman, J. Mitard, G. Hellings, A. Satta, V. Terzieva, L. Souriau, F. E. Leys, G. Pourtois, M. Houssa, G. Winderickx, E. Vrancken, S. Sioncke, K. Opsomer, G. Nicholas, M. Caymax, A. Stesmans, J. Van Steenbergen, P. W. Mertens, M. Meuris, and M. M. Heynsa, "Germanium MOSFET Devices: Advances in Materials Understanding, Process Development, and Electrical Performance," *J. Electrochem. Soc.* **155**, H552-H561 (2008).

[2] H.-Y. Yu, M. Ishibashi, J.-H. Park, M. Kobayashi, and K. C. Saraswat, "p-Channel Ge MOSFET by Selectively Heteroepitaxially Grown Ge on Si," *IEEE Electron Device Lett.* **30**, 675-677 (2009).

[3] C. L. Andre, J. A. Carlin, J. J. Boeckl, D. M. Wilt, M. A. Smith, A. J. Pitera, M. L. Lee, E. A. Fitzgerald, and S. A. Ringel, "Investigations of High-Performance GaAs Solar Cells Grown on Ge–$Si_{1-x}Ge_x$–Si Substrates," *IEEE Trans. Electron Dev.* **52**, 1055-1060 (2005).

[4] G. Masini, L. Colace, and G. Assanto, "Si based optoelectronics for communications," *Materials Science and Engineering B - Solid State Materials for Advanced Technology* **89**, 2-9 (2002).

[5] C. L. Andre, J. J. Boeckl, D. M. Wilt, A. J. Pitera, M. L. Lee, E. A. Fitzerald, B. M. Keyes, and S. A. Ringel, "Impact of dislocations on minority carrier electron and hole lifetimes in GaAs grown on metamorphic SiGe substrates," *Appl. Phys. Lett.* **84**, 3447-3449 (2004).

VO₂, a Metal-Insulator Transition Material for Nanoelectronic Applications

K. Martens[a,b], I. P. Radu[a,c], G. Rampelberg[d], J. Verbruggen[a,d], S. Cosemans[a], S. Mertens[a],
X. Shi[a], M. Schaekers[a], C. Huyghebaert[a], S. De Gendt[a,e], C. Detavernier[d], M. Heyns[a,f],
J.A. Kittl[a]

[a] IMEC, Kapeldreef 75, Leuven, Belgium
[b] ESAT Dept., KULeuven, Leuven, Belgium
[c] Physics Dept., K. U. Leuven, Leuven, Belgium
[d] Dept. of Solid State Sciences, UGent, Gent, Belgium
[e] Chemistry Dept. KULeuven
[f] Materials Engineering Dept., KULeuven

An overview is given of exploratory Metal-Insulator Transition material based nanoelectronics research at imec and UGent. The VO₂ thin film growth techniques used are elaborated. This considers thermal oxidation of vanadium metal and Atomic Layer Deposition (ALD). Fundamental device properties such as voltage driven switching and tunnel junction properties are discussed. Device concepts making use of MIT materials or VO₂ are elaborated.

Introduction

Materials showing a metal-to-insulator transition (MIT)[1] due to correlated electron behavior are being investigated for potential nano-electronic applications[2-8]. The binary transition metal oxide vanadium dioxide is of particular interest due to its near room temperature abrupt 68°C MIT[1]. Key VO₂ properties are its very fast transition[9], its full volume transition implying good scalability and reliability and its high on current in the metallic state. For memory applications, these properties entail significant advantages over one of the current leading candidates to replace flash memory, Resistive RAM.

Vanadium Dioxide Thin Film Growth Techniques

Two thin film growth techniques of VO₂ are described: the thermal oxidation of vanadium metal to vanadium dioxide and atomic layer deposition. Most techniques used in literature for VO₂ deposition are either RF or DC reactive PVD (Physical Vapor Deposition)[10,11], CVD (Chemical Vapor Deposition) [12], PLD (Pulsed Laser Deposition) [13,14,15] or sol gel [16]. Thermal oxidation and ALD are not commonly used to fabricate VO₂ test structures but are commonly used in CMOS technology mass production to grow gate dielectric films, the thinnest films for a MOSFET which are crucial for MOSFET functionality. The use of these techniques, especially ALD, should in term enable the production of highly scaled VO₂ films with high uniformity, conformality and yield compatible with existing fabrication infrastructure.

Vanadium Dioxide Thin Film Deposition Techniques: Thermal Oxidation

To obtain the VO₂ phase of vanadium by means of thermal oxidation it is key to use a suitable partial oxygen pressure and temperature range. The oxidation kinetics of thermal oxidation of vanadium to VO₂ were evaluated by performing oxidations in a reaction chamber with in-situ XRD capabilities [17]. Kinetic analysis was done starting from an 80nm V layer deposited using sputtering on a SiO_2/Si substrate. The oxidation of VO_2 to V_2O_5 was also evaluated. AFM was used to evaluate roughness of the films. The dependence of the magnitude of the resistance transition was studied in function of oxidation temperature and oxygen partial pressure.

Concerning the kinetics of oxidation of vanadium metal to VO_2 partial oxygen pressures of 0.1mbar to 0.8 mbar and temperatures from 488°C to 650°C were evaluated. In the 488°C to 544°C temperature range the growth was found to obey the Deal-Grove oxidation model[18]. The growth was observed to be planar and approximately linear indicating that diffusion is fast and that the reaction limits the oxidation process. The kinetics were observed to be independent of partial oxygen pressure.

When the oxidation temperature was raised above 600°C the nature of the process changed and became nucleation controlled and was described by a JMAK model (Johnson-Mehl-Avrami-Kolmogorov) [19-21]. The oxidation behavior in function of time shows the typical S-shaped curve of JMAK-type behavior. The time before the onset of oxidation, the incubation time, was found to be inversely proportional to the partial oxygen pressure and points to an oxygen absorption process before the onset of VO_2 formation. The Avrami coefficient, is first found to be 2 evolving into a value of 3 as the growth of VO_2 progresses. In the case of site saturation with a constant amount of nuclei in function of time this points to an evolution from 2D to 3D growth, and for a case of continuous nucleation this points to an evolution from 1D growth to 2D growth. By means of an SEM inspection in function of growth time it was found that the continuous nucleation occurs and hence a transition from 1D growth to 2D growth occurs. The oxidation of vanadium dioxide to vanadium pentoxide was found to obey the Deal-Grove model and showed planar growth and linear-parabolic kinetics. This entails that the process is both diffusion and reaction limited.

For the morphology of the films, the temperature plays a larger role than the oxygen pressure over the ranges investigated. Lower oxidation temperature translates into lower film roughness and lower particle size. For samples produced at higher temperatures typical rms roughness is of the order of 2-5nm for a 100nm-thick film as measured by atomic force microscopy.

The highest amplitude of the transition [22] in electrical sheet resistivity with temperature (nearly 4 orders of magnitude) was found for thermal VO_2 on crystalline Al_2O_3 grown at 475°C at 3mbar pO_2. From 475°C on the amplitude was found to decrease with increasing oxidation temperature at 3mbar pO_2. On amorphous Al_2O_3 and on SiO_2 the maximum amplitude was found to be around 3 orders of magnitude. On amorphous substrates VO_2 was textured in the [110] direction while on Al_2O_3 it was textured in the [020] direction. Generally the onset of V_2O_5 formation at higher oxygen pressures (>3mbar) reduces the amplitude of the transition.

Vanadium Dioxide Thin Film Deposition Techniques: Atomic Layer Deposition

Thin ALD vanadium dioxide films[23] were grown with a low-pressure ALD reactor in the temperature range 100°C-175°C on Si substrates with 100nm thermally grown SiO_2. TEMAV $\{V(NEtMe)_4\}$ was used as metal-organic vanadium precursor. The oxidation state of the vanadium in the TEMAV molecule is 4+, in contrast to other vanadium precursors like $VO[O(C_3H_7)^i]_3$ where it is 5+. The 4+ oxidation state could possibly be a driving force to obtain VO_2 instead of V_2O_5. As reactant gas ozone was used.

Since the main characteristic of ALD is the self-limiting nature of the surface reactions, it is important to check in which temperature range this really is the case. Therefore, a process was performed at several deposition temperatures consisting of 200 times 2s exposure with TEMAV while no reactant gas was used. For an ALD process this should not result in film growth due to the self-limiting nature of each ALD step and the need of interchanging oxidation and precursor steps. It was found using XRF that negligible decomposition of the precursor resulting in film growth occurred at or below 150°C, allowing an ALD process at those temperatures.

A second required property of an ALD process is the occurrence of saturation, meaning that increasing the duration of either the oxidizer or precursor ALD pulse does not result in an increased film growth rate. This is also a consequence of the self-limiting nature of ALD. A leveling off of the growth per cycle was observed to be present in saturation experiments (see figure 1a). At 150°C it was concluded that 2s TEMAV and 5s ozone resulted in a saturated ALD growth (figure 1a). Pump times were respectively 25s and 15s after the TEMAV and ozone pulse. The corresponding growth per cycle (GPC) is 0.76Å/c. The typical linearity of an ALD process was also shown to be present (see figure 1b).

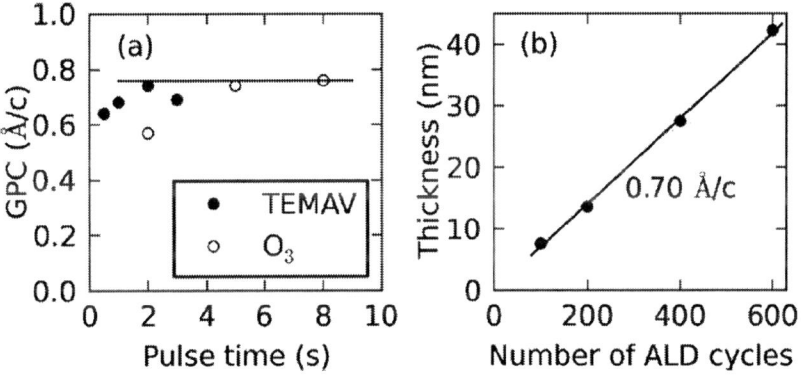

Figure 1. ALD characteristics for the process TEMAV – ozone: (a) Growth per cycle versus exposure times for TEMAV with a fixed ozone exposure of 5s (filled symbols) and for ozone with a fixed TEMAV exposure of 2s (open symbols); (b) Thickness of the deposited films as a function of the number of ALD cycles. [23]

The as grown ALD film was shown to be amorphous VO_2 with XPS measurements comparing the as grown ALD film, showing vanadium to be in a 4+ oxidation state, to a fully oxidized V_2O_5 films showing a 5+ oxidation state (see figure 2). A post anneal is done at a controlled oxygen pressure to convert the films into crystalline VO_2 as characterized with XRD and with sheet resistivity measurements as a function of temperature. The amplitude of transition over the metal-to-insulator transition was approximately 2 orders of magnitude. For a 42nm VO_2 film the RMS roughness was found to be 4.5nm using AFM.

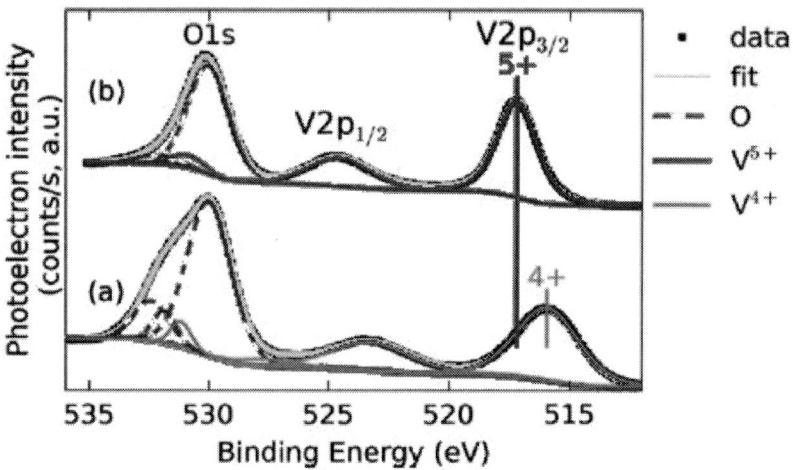

Fig.2 XPS of a film grown with TEMAV and ozone (a) O_2 plasma assisted ALD film shown to be vanadium pentoxide with XRD (b).[23]

Fundamental device properties

2-terminal vanadium dioxide devices in which the VO_2 is contacted by two electrodes patterned on top of a VO_2 film (see figure 3b) show a typical switching characteristic depicted in figure 3a. The distance between the electrodes is referred to as channel length. As the voltage is increased the characteristic shows an abrupt transition to a low resistance regime. When the voltage is decreased subsequently a transition occurs back to the original high resistance regime at a lower voltage. The switching is symmetric about zero and is volatile. This characteristic can be explained by Joule heating [24,25,26].

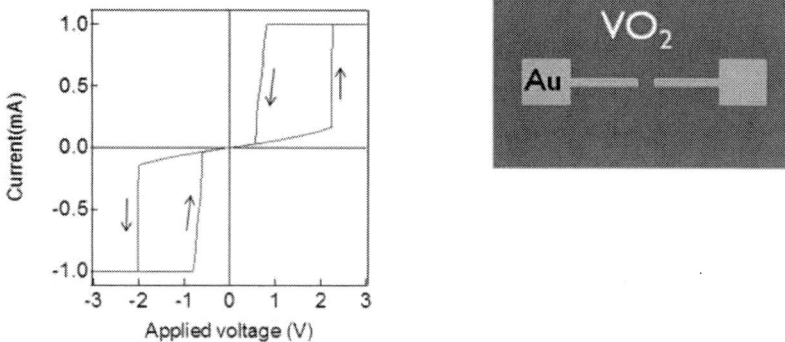

Figure 3a.   Current-Voltage characteristic   Figure 3b.   Top view of VO_2 lateral 2-

of a lateral 0.1x10μm VO$_2$ device.   terminal device.

The voltages and fields at which the device switches on (high to low resistance state) and off (vice versa) were modeled. The on and off fields are, for larger channel lengths, more or less constant with field (see figure 8) or equivalently the on voltage was observed to be approximately proportional to channel length. A first order model can explain the behavior for longer channel lengths (>1μm). In steady state the heat drained from the VO$_2$ must equal the heating power supplied by Joule heating. At the transition temperature this will determine the on voltage: $V_{on}^2/R_L = G_{th}(T_t - T_{amb})$ in which V_{on} is the on voltage, R_L is the high resistance in the low temperature state of VO$_2$, G_{th} is the thermal conductance draining heat from the VO$_2$, T_t is the transition temperature and T_{amb} the ambient temperature. For the off voltage this becomes: $V_{off}^2/R_H = G_{th}(T_t - T_{amb})$ in which R_H is the low resistance in the high temperature state. So $V_{on} = \sqrt{(R_L G_{th}(T_t - T_{amb}))}$ and an analogous expression is valid for V_{off}. The off voltage is lower than the on voltage because a low resistance can be kept heated to a higher temperature with the same voltage. Since $R \propto L$ and $G_{th} \propto L$ (assuming heat draining through substrate only) this means that $V_{on} \propto L$. This is for the first order model. The linear dependence is deviated from for shorter device lengths. This can be explained by a second order model which includes contact resistivity and heat draining through the contacts. This model fits the experimental data well with the used material parameters such as substrate heat conductivity, VO$_2$ resistivity and contact resistance in good agreement with expected values. The structures of channel length 0.1 - 2μm and width 10μm match well to this straightforward *uniform* Joule heating model, most likely due to the structure's small size. Cycling behavior [25, 26] was measured on the lateral VO$_2$ devices as depicted in figure 4a. Devices were repeatedly switched between the highly and lowly conductive state by applying appropriate voltages. The amount of cycles reached without loss of on and off state properties reached 10Gcycles.

In a different experiment isothermal retention [22] was measured on thin films of VO$_2$ (not devices) with four point probe measurements at 68C. It can be concluded that the on and off state can be maintained up to 10 years using extrapolation (see figure 4).

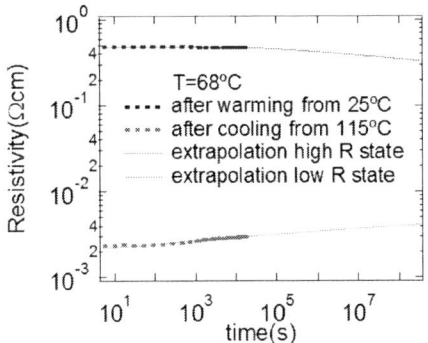

Figure 4. Isothermal retention measured on VO$_2$ thin films.[22]

We show that the VO$_2$ MIT manifests itself in an abrupt interfacial transition of at least two orders of magnitude in VO$_2$ tunnel junctions formed by a thin barrier layer of native V$_6$O$_{13}$ on VO$_2$ [27, 28]. Characterization was done with TEM, XPS, specific contact resistivity measurements of CTLM (Circular Transmission Line Method) structures (see figure 5) and VO_2 work function extractions. Wentzel-Kramers-Brillouin modeling explains the phenomenon in detail. The viability is shown of using tunnel junctions to scalably modify the 'off' resistance of MIT components, which is generally too low for nano-electronics.

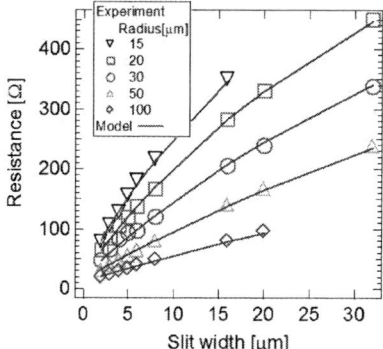

Figure 5. Resistance of VO_2 CTLM structures as a function of geometrical parameters used to extract contact resistivity.

The interfacial resistivity or contact resistivity of VO_2-barrier-metal structures was measured by using CTLM structures for a series of different metals. We observed a relative insensitivity of the extracted contact resistivity to the metal work function. Another observation is the large change in contact resistivity across the metal insulator transition at 68°C. The temperature dependence was measured in more detail showing clearly that the abrupt metal-insulator transition manifests itself in the interfacial resistivity. This effect was explained by using a WKB tunneling model which shows that the transition in interfacial resistivity is induced by a change in charge carriers in the VO_2. Three different configurations were modeled of which the 'thin high' barrier (5Å-1nm thick dielectric with conduction band offset > 1.5eV) fitted the observations best. When the VO_2 becomes metallic much more charge carriers are present available for tunneling. Hence the strongly increased tunneling current through the barrier into the metal.

Further MOS stacks consisting of VO_2-high-κ/metal with VO_2 as the semiconductor were studied. The MOS capacitor electrical properties are analyzed such as the gate current and capacitance behavior with special regard to the MIT.

Device Concepts

Several device concepts were developed making use of metal-insulator transition materials. It was studied how devices could be created based on the signature characteristic of MIT materials such as VO_2 which show a sudden transition in electrical resistivity with temperature. This temperature behavior translates itself into the typical voltage dependence of the VO_2 devices which show a volatile switching loop (see figure 3).

Selector device: A first concept developed was a selector device based on VO_2. In a memory array, a selector device is required to activate only the cells of the accessed word. For the read operation, the selector must suppress the leakage current of the non-active cells. For the write operation, the selector must additionally absorb a sufficient part of the applied voltage to avoid disturbs of the non-selected cells.

The typical volatile voltage-switching behavior of VO_2 is very desirable to achieve these goals, hence VO_2 is a very promising selector concept. Three essential issues have to be dealt with to ensure VO_2 can serve as a selector device. Firstly, the VO_2 by itself is too conductive to serve as a selector. Secondly, a VO_2 element fulfilling selector specifications will be too resistive for self-heating. Thirdly, the transition temperature of the VO_2 needs to be raised sufficiently to make sure the selector is compatible with the operating temperature range specified for the chip. Alternatively, the voltage of the heater can be adapted to the ambient temperature.

Non-volatile memory: MIT devices have a bistable temperature range in which both the high and low resistivity phases are stable. If the bistable temperature range of an MIT material covers the entire standard operating temperature range for the chip, it can be used to create a non-volatile memory. The main challenge for this type of memory is hence to obtain a material with an as wide as possible bistable temperature range around room temperature to evolve the device toward compatibility with standard operating temperature ranges. Alternatively applications can be restricted to stable temperature environments.

A third concept involves complementary logic in which Peltier devices of opposite polarity are thermally connected to MIT elements. It is shown that logic can be created giving rise to inverters, nor and nand gates.

Conclusions

In summary, an overview is given of Metal-Insulator Transition material based technology research. Thin film growth techniques of VO_2 are covered: thermal oxidation of vanadium metal and atomic layer deposition. Fundamental device properties are investigated such as voltage induced switching in VO_2 and interfacial resistivity transitions induced by the VO_2. Device concepts based on MIT materials are elaborated.

Acknowledgments

KM and IPR acknowledge financial support from FWO Vlaanderen. IPR acknowledges funding through a Marie Curie Reintegration Grant. C.D. acknowledges the European Research Council for funding (grant agreement n° 239865). We thank Nico Jossart, A. Stesmans, H. Tielens, J. Steenbergen, W. Magnus, W. Vandenberghe, J. Musschoot, C. Huyghebaert, and G. Groeseneken.

References

[1] F. J. Morin, *Phys. Rev. Lett.* **3**, 34 (1959).
[2] Newns et al. APL 73, p.780 (1998)
[3] Hong et al. , APL 86, p. 142501 (2005).
[4] M.-J. Lee, Y. Park, D.-S. Suh, E.-H. Lee, S. Seo, D.-C. Kim, R. Jung, B.-S. Kang, S.-E. Ahn, C. Lee, et al., Advanced Materials **19**, 3919 (2007).
[6] Son et al., IEEE EDL 32 (1), p. 1579 (2011)

[7] H. T. Kim, B. G. Chae, D. H. Youn, S. L. Maeng, G. Kim, K. Y. Kang, and Y. S Lim, *New J. Phys.* **6**, 52 (2004)

[8] C. Ko and S. Ramanathan, Applied Physics Letters **93**, 3050464, 2008.

[9] A. Cavalleri, T. Dekorsy, H. H. W. Chong, J. C. Kieffer, and R. W. Schoenlein, Phys. Rev. B **70**, 161102, 2004.

[10] P. Jin, K. Yoshimura, and S. Tanemura, *J. Vac. Sci. Technol. A* **15**, 1113 (1997).

[11] D. Ruzmetov, K. T. Zawilski, V. Narayanamurti, and S. Ramanathan, *J. Appl. Phys.* 102, 113715 (2007).

[12]M. B. Sahana, G. N. Subbanna, and S. A. Shivashankar, *J. Appl. Phys.* 92, 6495 (2002).

[13] D. H. Kim and H. S. Kwok, *Appl. Phys. Lett.* **65**, 3188 (1994).

[14] H. S. Choi, J. S. Ahn, J. H. Jung, and T. W. Noh, *Phys. Rev. B* **54**, 4621 (1996).

[15] M. Soltani, M. Chaker, E. Haddad, R. V. Kruzelecky, and D. Nikanpour, *J. Vac. Sci. Technol. A* **22**, 859 (2004).

[16] F. Beteille et J. Livage, *J. Sol-Gel Sci. Tech.* **13**, 915 (1998).

[17] G. Rampelberg et al. to be published

[18] A. B.Deal, vol. 36, no. 1, pp. 3770–3779, 1965.

[19] Kolmogorov, Bull. Acad. Sci. USSR, no. 1, p. 355, 1937.

[20] P.A. Mehl, W.A. Johnson, Trans. AIMME, no. 135, p. 416, 1939.

[21] M. Avrami, The Journal of Chemical Physics, vol. 7, no. 12, pp. 1103–1112, 1939.

[22] I.P. Radu, K. Martens, S. Mertens, C. Adelmann, X. Shi, H. Tielens, M. Schaekers, G. Pourtois, S. Van Elshocht, S. De Gendt, M. Heyns , J. A. Kittl, , ECS Spring Meeting 2011, Montreal, Canada, May 2011

[23] Rampelberg Geert, Schaekers Marc, Martens Koen, Xie Qi, Deduytsche Davy, De Schutter Bob, Blasco Nicolas, Kittl Jorge, Detavernier Christophe, Appl. Phys. Lett. 98, 162902, 2011

[24] J. Duchene, M.M. Terraillon, M. Pailly, G.B. Adam, IEEE TED Vol. ed-18, no. 12, dec. 1971

[25] Iuliana P. Radu, K. Martens, B. Govoreanu, S. Mertens, X. Shi, M. Cantoro, M. Schaekers, M. Jurczak, S. De Gendt, A. Stesmans, M. Heyns, J. A. Kittl, Solid State Devices Meeting, Nagoya, Japan, September 2011

[26] Iuliana P. Radu, K. Martens, B. Govoreanu, S. Mertens, X. Shi, M. Cantoro, M. Schaekers, M. Jurczak, S. De Gendt, A. Stesmans, M. Heyns, and J. A. Kittl, submitted to *Applied Physics Letters, 2012*

[27] K. Martens, I.P. Radu, S. Mertens, X. Shi, M. Schaekers, H. Tielens, C.Huyghebaert, S. Degendt, M. Jurczak, V. Afanas'ev, M. Heyns, J.A. Kittl, *IEEE Semiconductor Interface Specialists Conference*, Arlington, USA, Dec. 2011

[28] K. Martens, I.P. Radu, S. Mertens, X. Shi, P. Favia, H. Bender, T. Conard, M. Schaekers, M. Jurczak, S. De Gendt, V. Afanas'ev, M. Heyns, and J.A. Kittl, submitted to *Applied Physics Letters, 2012*

ECS Transactions, 45 (4) 159-167 (2012)
10.1149/1.3700466 ©The Electrochemical Society

Demonstration of single crystal GaAs layers on CTE-matched substrates by the Smart Cut[TM] technology

T. Jouanneau[a], Y. Bogumilowicz[a], P. Gergaud[a], V. Delaye[a], J-P. Barnes[a], V. Klinger[b], F.Dimroth[b], A. Tauzin[a], B. Ghyselen[c], and V. Carron[a]

[a] CEA, LETI, MINATEC Campus, 17 rue des Martyrs, 38054 GRENOBLE Cedex 9, France
[b] Fraunhofer Institut for Solar Energy Systems, Heidenhofstrasse 2, 79110 Freiburg, Germany
[c] SOITEC S.A., Parc technologique des fontaines, 38190 Bernin, France

Templates made of a thin single crystal GaAs layer on CTE-matched substrate (sapphire) have been realized using the Smart Cut[TM] technology. These templates can withstand high processing temperatures thanks to the CTE matching between the GaAs thin film and its support, and therefore can be used for many applications since they require no specific restriction concerning thermal treatments. The GaAs templates have been compared to conventional bulk GaAs substrates. TEM images and XRD spectra show similar crystalline quality. AFM measurements show a similar surface microroughness. Photoluminescence of a AlGaAs double heterostructure grown by MOCVD on the GaAs template shows the same intensity as a reference structure on bulk GaAs. Therefore, the GaAs templates can replace GaAs bulk substrates in various domains such as photonics devices (solar cell, laser) or high frequency electronics.

Introduction

Gallium arsenide is one of the most popular compound semiconductors used in the industry. Indeed, it has very interesting properties such as a very high electron mobility of 8200 cm²/Vs (1), which enable transistor switching frequencies of up to 500GHz (2). Moreover, the higher GaAs bandgap of 1.42eV makes it less sensitive to thermal noise than silicon. It is therefore widely used in telecommunication devices, where high frequency operation is required. However, the lack of a stable GaAs oxide has limited its use, especially in the microelectronics industry where silicon clearly dominates. Unlike silicon, GaAs has a direct bandgap. It is then a very interesting material for optoelectronic applications such as lasers, LEDs and high efficiency solar cells (3). There is however a limitation related to the GaAs substrate: its cost represents a non negligible part of such optoelectronic devices (especially for solar cells where there is strong constraints on prices). In addition, the major part of GaAs in the substrate has no active role in the device besides providing mechanical support. Therefore, rare and expensive materials are wasted where they could be replaced by a lower cost alternative. It is then interesting in term of material needs to use GaAs templates consisting of a thin (<1 μm) layer of GaAs on a mechanical support made using the Smart Cut[TM] technology instead of bulk (>200 μm) GaAs substrates.

Researchers have so tried to adapt the Smart Cut[TM] technology for the transfer of GaAs thin films onto a cheaper substrate. For example, Jalaguier *et al.* reported the first transfer of a GaAs layer onto a silicon substrate. This transfer was either done using a silicon oxide bonding (4) or a metallic bonding (5). Radu *et al.* (6) also reported a transfer of GaAs on silicon, using a Spin On Glass layer for the bonding. Devices like solar cells using this alternative substrate approach have been realized by Schöne *et al.* (7) on a GaAs on silicon template. However, because of the strong CTE (Coefficient of Thermal Expansion) mismatch between silicon and GaAs (2.6×10^{-6} / K (8) and 6.0×10^{-6} / K (1) respectively at room temperature), cracks were induced in the active layer when cooling down the substrate after the epitaxy.

In order to overcome the limitations of previous attempts of providing GaAs thin films, we propose a new stack consisting of thin GaAs layers on CTE-matched substrates using the Smart Cut[TM] technology. A CTE matched substrate avoids putting the thin GaAs film under high stress when performing thermal treatments and therefore avoids the undesirable effects observed when using GaAs on silicon. Possible candidates are compared in Figure 1 where CTE of GaAs, Ge (9), Si and c-plane sapphire (10) are plotted in function of temperature. Ge is nearly identical to GaAs. The CTE of Sapphire is lower than GaAs one below 200°C and becomes similar after 300°C. Silicon expansion remains clearly lower over the whole temperature range. Therefore, GaAs, Ge and Sapphire are considered "CTE-matched".

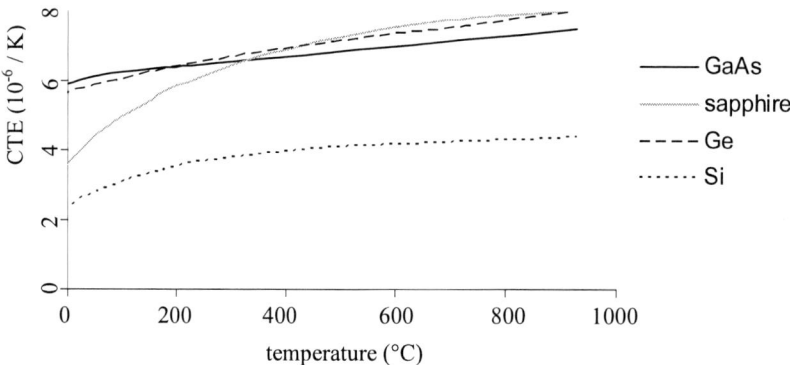

Figure 1 : Evolution of Coefficient of Thermal Expansion (CTE) with temperature. GaAs (100) is compared with Si (100), Ge (100) and sapphire (c-plane).

Among the CTE matched materials, using sapphire allows reducing the amount of semiconductor material necessary for the substrate. Sapphire has other advantages like it provides a strong electrostatic decoupling between the active layer and the substrate, which would be interesting for making RF devices. In addition, sapphire is composed of single-crystal Al_2O_3, which components are abundant on earth. From a practical point of view, it can be treated in a microelectronic fabrication plant without cross contamination risks. It is a strong material that hardly breaks and finally it is transparent in the visible range, which can be interesting for optoelectronic applications.

The Smart Cut™ technology allows to obtain thin (<1μm) GaAs layers on sapphire and recycle the donor GaAs substrate to reduce rare and expensive material consumption. We will present the Smart Cut™ technology used to make GaAs templates. Such GaAs templates have then been characterized in order to assess their material quality and in particular compare them to bulk GaAs substrates. AFM, TEM and XRD results will be shown. III-V epitaxial layers have then been grown by Metal-Organic Chemical Vapor Deposition (MOCVD) in order to test the suitability of the GaAs templates to epitaxy. Such epi-layers have been characterized by Photoluminescence. The GaAs templates realized on sapphire substrates will also be further mentioned using the acronym GaAsOS (for GaAs On Sapphire).

Experimental details

Starting material

The starting materials used to realize the GaAsOS templates are 100mm diameter bulk gallium arsenide and sapphire substrates. The bulk GaAs substrates are (001) surface oriented with an off cut of 6° towards [111] crystalline direction. Sapphire surface orientation is (0001) corresponding to the c-plane.

GaAsOS template realization

GaAsOS templates are realized using the Smart Cut™ technology, whose basic process steps are summarized in Figure 2. This process, which is used industrially to produce silicon-on-insulator substrates, allows recycling of the GaAs donor wafer at the end of the process, so that only the needed quantity of GaAs is consumed.

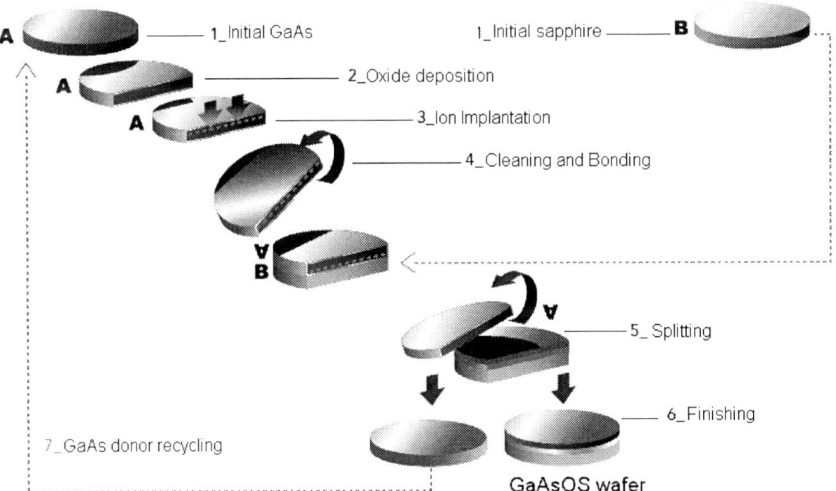

Figure 2 : Schematic of the basic process steps involved in the Smart Cut™ technology.

First of all, bonding layers are deposited on the two initial substrates. Then hydrogen is implanted in the GaAs substrate at a dose comprised between 10^{16} and 10^{17} H/cm² in order to create a weakened layer. GaAs and sapphire substrates are bonded by direct bonding. After a low temperature (<450°C) annealing, splitting occurs in the hydrogen implanted zone, allowing detachment of a thin single crystal film from the GaAs donor substrate. The resulting GaAsOS template is then polished by Chemical Mechanical Polishing (CMP) and cleaned. The surface is chemically prepared to obtain *epi-ready* surfaces. Such substrates are then used for epitaxial purposes. To simulate the effect of an epitaxy pre-bake treatment on GaAs layer characteristics, some of our GaAsOS templates were taken just before epitaxy and annealed at high temperature (>450°C).

GaAsOS template characterization

Transmission electron microscopy (TEM) images were done using a 200 kV Tecnai G² F20 STWIN tool from FEI. Samples were prepared beforehand using a Focused Ion Beam–Scanning Electron Microscope dual beam FEI Strata 400 apparatus. Roughness was measured by Atomic Force Microscopy (AFM) in tapping mode on a Digital Instrument Dimension 3100 system. High resolution X-Ray diffraction (HR-XRD) spectra have been obtained using an Xpert MRD PANalytical diffractometer. Hydrogen profiling was done by Secondary Ion Mass Spectroscopy (SIMS) on an ION TOF ToF-SIMS V instrument with 2keV Cs$^+$ sputtering.

Epitaxy tests

A AlGaAs double heterostructure (Figure 3) has been grown on the GaAsOS templates to test their suitability to epitaxy. Indeed, good crystalline quality epilayers will be achieved only if the GaAsOS templates exhibit a smooth surface and a good passivating layer that can be desorbed thermally. Since this process is conducted at relatively high temperatures (600°C-700°C), it is also a good thermal stability test for the GaAsOS templates. The same structure was grown in parallel on a reference bulk GaAs wafer to compare grown layer quality. Epitaxy has been realized by means of MOCVD using an AIX2600-G3 reactor.

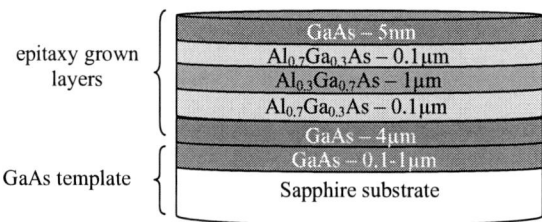

Figure 3 : GaAlAs double heterostructure grown on GaAsOS template.

Arsine was used for the group-V growth. TMGa and TMAl were the group-III precursors. Doping source was SiH$_4$. Typical growth temperatures range between 600 - 700°C and reactor pressures between 50 - 100 mbar. Photoluminescence (PL) intensity mapping have been carried out at room temperature using a 514 nm wavelength argon-laser.

Results and discussion

<u>General aspect</u>

The general aspect of the GaAsOS template after polishing can be seen in Figure 4. It shows that the transferred thin GaAs film covers almost the whole surface of the holding substrate, at the exception of the periphery (~2mm large). This is typical of an SOI-like wafer since the substrates are beveled at the edge. A few small round shaped areas where the GaAs thin film was not transferred can be observed as well. The origin of such areas is the presence of particles at some critical steps during the realization, such as before direct bonding or before implantation. One can notice different colours on the wafer, which are interferences due to the finite thickness of the GaAs layer. The color shift shows a thickness variation, which is consistent with spectroscopic ellipsometry measurements of the GaAs thickness. Thickness uniformity is 0.33 +/-0.03 μm, and GaAs thin film is slightly thicker at the center than at the edge.

Figure 4 : General aspect of a 100 mm diameter GaAsOS template after polishing.

<u>Morphological characterization</u>

Microroughness of the GaAs surface of the template was probed after polishing by performing AFM on a 5 x 5μm² zone. The calculated root mean square (RMS) value is 0.23 nm. Using the same measurement conditions on a bulk GaAs substrate, we measured a RMS roughness of 0.15 nm. The GaAsOS template therefore has a similar RMS roughness than a bulk GaAs substrate, although slightly higher. Such a low RMS roughness value for the GaAsOS template is nevertheless compatible with process steps requiring smooth surfaces, such as epitaxy or photolithography.

Transmission Electron Microscope cross-sections images were done after polishing of the GaAs surface to evaluate the crystalline quality of the GaAs layer. One of these cross sections is shown in Figure 5. Defect density on GaAs layer is low enough so as no defect is detected by this technique. High Resolution-TEM images (not shown here) have also revealed that the layer is a single-crystal on the observed sample.

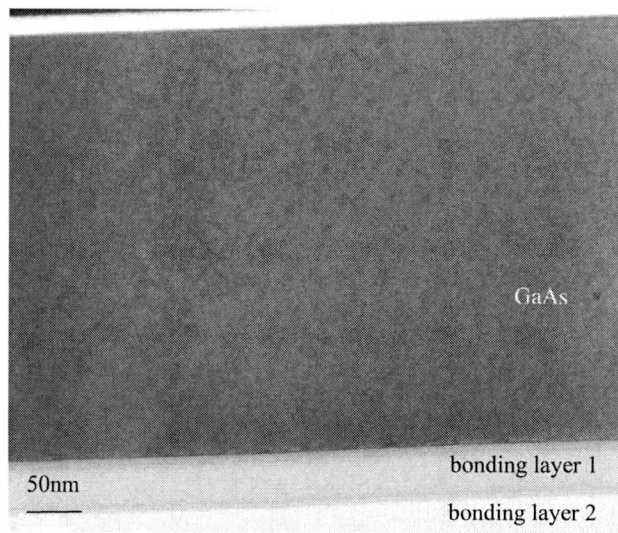

Figure 5 : TEM cross section of GaAsOS template after polishing.

Physicochemical characterizations

Figure 6 shows HR-XRD spectra in theta-2theta configuration on the (004) planes of the single-crystal GaAs layer of the GaAsOS template after CMP, and after the final high temperature anneal. These layers are compared with a bulk GaAs reference, whose peak Full Width at Half Maximum (FWHM) is 10.08 arcsec. This is the range of values that can be measured with this experimental setup on high-quality single-crystal Silicon wafers.

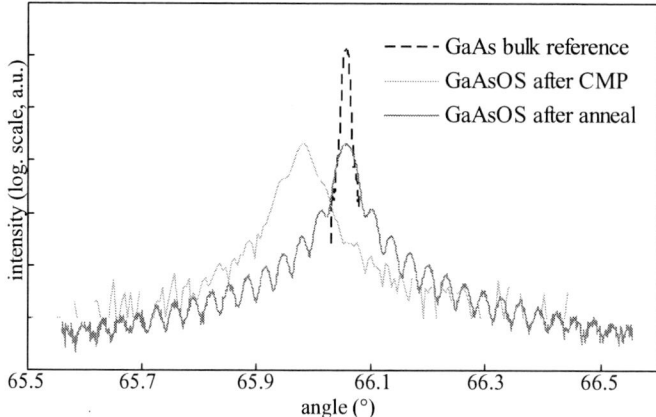

Figure 6 : High resolution XRD spectra of GaAsOS templates at different stages of their realization and of a GaAs bulk reference. (004) GaAs reflection is measured in theta-2theta configuration.

The (004) reflection corresponding to the GaAsOS template just after CMP is shifted by 266.4 arcsec towards lower angular values. This is characteristic of an in-plane compressive stress of 123MPa, that is due to the residual presence of implanted hydrogen, as we will see further in the SIMS H profiling. The higher FWHM (275.0 arcsec) is due to the finite thickness of the GaAs layer.

After the high temperature anneal, the GaAsOS template diffraction peak gets back to the same position of the bulk reference, which means that the stress of the layer has been relieved. We can observe interference fringes that usually appear on layer with a finite thickness and with sharp interfaces. Indeed, the XRD spectrum fits well to a simulated spectrum of a single crystal defect-free 330 nm thick GaAs layer.

According to a preferred option of the Smart Cut[TM] technology, H implantation has been used to induce the separation of a GaAs single crystal layer from a donor substrate. SIMS has been used in this work to characterize and track hydrogen traces in layers obtained by this method. Figure 7 shows these measurements. We can observe the hydrogen concentration profiles measured on GaAsOS templates after the CMP of the GaAs surface and after the high temperature anneal.

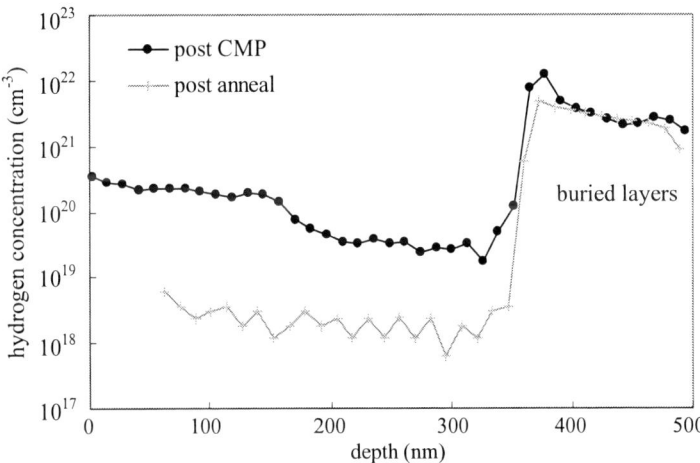

Figure 7 : SIMS H depth profiling of GaAsOS template after CMP and after annealing. Concentration values are valid only in GaAs layer. Post-anneal curve is shifted to obtain GaAs / buried layers interface at the same depth.

As we would expect from the XRD measurements, hydrogen is detected after polishing and its concentration is going down with depth. This is expected as we are moving away form the cleaving plane, which was located near the maximum hydrogen concentration region after implantation. After the high temperature anneal, hydrogen concentration drops down to the noise level, which is consistent with the XRD data that shows that the compressive stress disappeared.

Epitaxy tests

One good method of assessing the performance of the GaAsOS template is to realize an epitaxy on top of it and measure the quality of the grown structures by mean of photoluminescence. Indeed, the intensity of the emitted light of a semiconductor, when optically pumped, is a good indicator of the radiative recombination processes taking place. Defects present in the semiconductor material usually lead to non-radiative recombination processes, and therefore a high PL intensity is a good indicator for a good crystalline quality of the epitaxy layers.

Room temperature PL results can be seen in Figure 8. It shows the PL intensity of the same double AlGaAs heterostructure grown either on a GaAs substrate (left-hand side) and a GaAsOS template (on the right-hand side). The PL intensity is measured on an arbitrary scale compared to an internal reference. This is why intensities higher than 100% are sometimes achieved.

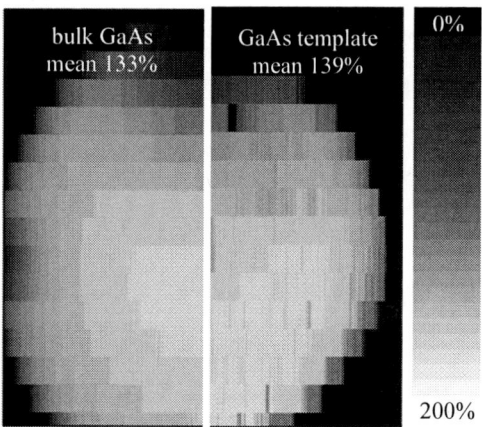

Figure 8 : Photoluminescence intensity mappings on GaAlAs double heterostructures realized on GaAsOS template (right image) and bulk GaAs reference wafer(left image). The intensity percentages shown refer to an internal reference and therefore can be higher than 100%.

First of all, we observe that the use of a GaAsOS template does not degrade at all the PL intensity: the mean value (139%) is even slightly higher on the GaAsOS template than on the reference bulk GaAs substrate (133%). PL emission wavelength is also almost the same: 670.8nm for the structure grown on GaAsOS template while the same structure emits at 671.2 nm when grown on bulk GaAs reference. No difference can be made between a GaAsOS template and a bulk GaAs substrate from the PL point of view. Optical microscope images have also shown that after the epitaxy, no cracks are observed in the several micrometers thick grown layer, which is a strong improvement with respect to previous experiments when templates of GaAs on silicon were used (7). Using a CTE matched substrates makes the template highly suitable for epitaxial growth.

Conclusion

We have fabricated GaAsOS (GaAs on sapphire) templates in order to offer alternatives to bulk GaAs substrates for applications where only a thin (<1 μm) layer of GaAs is necessary for device operation. The GaAs layer is stress-free and exhibits a very good crystalline quality, as measured by TEM and XRD. Surface microroughness of our GaAsOS templates, as measured by AFM, is similar to the one obtained on bulk GaAs. An AlGaAs photoluminescence test structure has been grown on both a GaAsOS template and a bulk GaAs reference, showing the same photoluminescence intensity in both cases. This proves the high quality of the epitaxy layers on the new GaAsOS template. Finally, the GaAsOS template did not show any cracks in the grown layer, validating the ability to withstand high processing temperatures. This is a strong improvement with respect to previous attempts of providing GaAs films on a handling substrate. In conclusion, GaAsOS templates appear as a good alternative to GaAs bulk wafers, whenever only a thin GaAs layer is required for the application.

Acknowledgments

This work was partly supported by the French ANR and German BMBF through the Inter-Carnot Fraunhofer program "SolarBond" project. Christophe Wyon is gratefully acknowledged for his support. Smart Cut^{TM} is a trade mark from SOITEC S.A.

References

1. M. R. Brozel and G. E. Stillman, *Properties of Gallium Arsenide,* INSPEC, London (1996).
2. A. Tessman, A. Leuther, V. Hurm, I. Kallfass, H. Massler, M. Kuri, M. Riessle, M. Zink, R. Loesch, M. Seelmann-Eggebert, et. al, *IEEE J. of Solid-State Circuits, 46* (**10**), pp. 2193-2202 (2011).
3. H. Cotal, C. Fetzer, J. Boisvert, G. Kinsey, R. King, P. Hebert, H. Yoon, N. Karam, et. al., *Energy and Environmental Science* 2 (**2**), pp. 174-192 (2009).
4. E. Jalaguier, B. Aspar, S. Pocas, J.F. Michaud, M. Zussy, A.M. Papon, M. Bruel, *Electron. Lett.,* 34 (**4**), pp. 408-409 (1998).
5. B. Aspar, E. Jalaguier, A. Mas, C. Locatelli, O. Rayssac, H. Moriceau, S. Pocas, A.M. Papon, J.F. Michaud, M. Bruel, *Electron. Lett.* 35 (**12**), pp. 1024-1025 (1999).
6. I. Radu, I. Szafraniak, R. Scholz, M. Alexe, U. Gösele, *J. of Applied Physics,* 94 (**12**), pp. 7820-7825 (2003).
7. J. Schöne, F. Dimroth, A.W. Bett, A. Tauzin, C. Jaussaud, J.C. Roussin, *Conf. records of 2006 4th IEEE WCPEC-4* (**1**), pp. 776-779 (2007).
8. R. Hull, *Properties of Crystalline Silicon*, INSPEC, London (1999).
9. G. Grimvall, *Thermophysical Properties of Materials,* Vol 13, Elsevier, Amsterdam (1999).
10. E.R. Dobrovinskaya, L.A. Lytvynov, V. Pischik, *Sappphire, Material, Manufacturing, Applications*, Springer, New York (2009).

ECS Transactions, 45 (4) 169-180 (2012)
10.1149/1.3700467 ©The Electrochemical Society

Characterization of Rapid Melt Growth (RMG) Process for High Quality Thin Film Germanium on Insulator

N. Zainal[a,c], S. J. N. Mitchell[a], D. W. McNeill[a], M. F. Bain[a], B. M. Armstrong [a], P. T. Baine [a], D. Adley [b], T. S. Perova [b]

[a] Northern Ireland Semiconductor Research Centre, School of Electronics, Electrical Engineering & Computer Science, Queen's University, Belfast, BT9 5AH, UK
[b] Department of Electronic and Electrical Engineering, University of Dublin, Dublin 2, Ireland
[c] Department of Electrical and Electronics Engineering, Universiti Tun Hussein Onn Malaysia, 86400 Parit Raja, Batu Pahat, Johor, Malaysia

> Germanium is one of the most promising materials for high performance infra-red photovoltaic devices. High quality single-crystal germanium on insulator structures can be produced by a Rapid Melt Growth process. Experiments show that thin-film germanium deposited by physical vapor deposition provides better quality in comparison with chemical vapor deposition. The longitudinal optical Ge-Ge peak in Raman spectrum is shifted from the expected 300.2 cm^{-1} position due to tensile stress resulting from the thermal expansion differences of the materials. The importance of silicon in the rapid melt process is confirmed by the fact that germanium films on sapphire substrates yielded polycrystalline structure. Films produced at high temperature (980 °C) show Ge-Ge Raman peak with linewidth of 3.3 cm^{-1} indicating good crystalline quality, comparable to bulk germanium (3.2 cm^{-1}), and thus demonstrating the potential to produce low cost high quality germanium films.

Introduction

Semiconductor based photovoltaic technology has been developed for more than 20 years using a range of techniques and materials with first invention being silicon based. Research has aimed to improve the energy conversion efficiency and reduce the manufacturing cost [1, 2]. At present, multi-junction solar cells show the highest performance, but have complex and expensive manufacturing processes [3, 4]. By developing a deposition and growth technique with good material electrical properties, a low cost and high quality solar cell can be produced. Germanium has been chosen because it has excellent electrical properties. With a band gap of Eg=0.66 eV [5] energy from the infra red region of the solar or thermal spectrum can be absorbed and converted into electrical energy. The issues of availability and cost can be overcome by using thin-films instead of thick wafers of germanium. Solid-state structures can be produced using low cost thin-film deposition techniques combined the rapid melt growth (RMG) process. High-quality single crystal thin-films of germanium produced by RMG can provide a germanium on insulator (GeOI) structure. Initial experiments on integration of GeOI with

169

silicon and sapphire substrates have demonstrated that good quality germanium crystalline structure can be obtained as shown by micro-Raman spectroscopy.

Materials and Methods

In this work, two techniques have been used for deposition of germanium, namely physical vapor deposition (PVD) and chemical vapor deposition (CVD). The process parameters are also varied, aiming for an optimal germanium crystalline structure and to verify the lateral growth of germanium on insulator. The parameters include germanium thickness, crucible materials (silicon dioxide, hafnium dioxide), anneal temperature and substrate.

The rapid melt growth (RMG) process uses a micro-crucible to hold the molten germanium [6, 7]. The crucible is formed on the surface of an oxidized silicon wafer with the germanium contacting the silicon in a seed window. In the fabrication process for PVD samples, the germanium is deposited directly on top of the oxide. The germanium chemical vapor deposition process, however, is selective to silicon surfaces and thus requires a thin silicon layer to be deposited on the oxide prior to the germanium deposition. The deposited germanium is then patterned into narrow stripe features with 3 to 4μm width and 60 to 400μm lengths. These stripes are then covered with a capping layer of low temperature oxide to form the micro-crucible as shown in figure 1(a). Rapid thermal annealing (RTA) is applied to heat the germanium above its melting point (938 °C) for a few seconds. After annealing, the germanium cools and crystallizes starting from the silicon seed and continuing laterally along the crucible. The seed region is expected to be defective because of the lattice mismatch between silicon and germanium. The dislocations will extend out from the seed region but terminate at the micro-crucible walls, and thus good crystalline quality can be obtained beyond this region. A surface view of the structure with seed region of 60μm length and 3 to 4μm width germanium stripes is shown in the Scanning Electron Microscopy image, figure 1(b). Straight, smooth stripes with no delamination are observed before heat treatment (anneal) of the sample. During the RTA, the crucible can become strained due to the germanium balling phenomenon [8]. In our work the capping layer was reinforced using a polycrystalline silicon layer which was found to be essential in minimizing germanium balling and delamination [9]. The quality of germanium crystalline structure is examined by micro-Raman spectroscopy with Scanning Electron Microscopy (SEM), used for imaging.

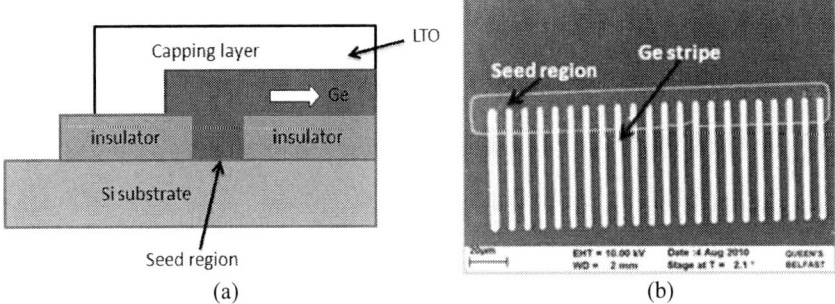

(a)   (b)

Figure 1. (a) Schematic diagram (not to scale) of rapid melt growth process with low temperature oxide (LTO) as a capping layer. (b) Micrograph of germanium stripes before annealing process.

Results and Discussions. In the case of a capping layer comprising only low temperature oxide (PECVD oxide), significant cracks and delamination of the stripes are observed and crucible integrity is compromised as shown in figure 2. This is because when heat treatment is applied above germanium melting point (938 °C), it tends to form into ball shapes (balling phenomenon) [8] causing the crucible to become strained. The issue is more severe with wide stripe since micro-crucible cap is wider and hence less rigid.

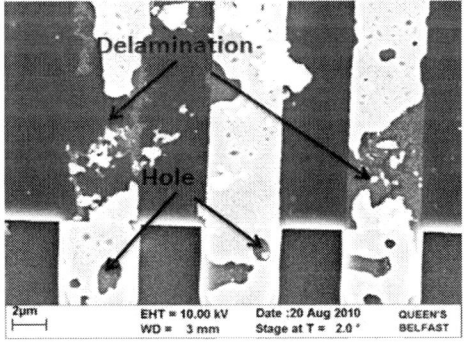

Figure 2. Micrograph showing cracks and delamination of germanium stripe with only PECVD oxide cap after annealed at 942 °C for 1 second.

During the heat treatment to melt the germanium, the oxide capping layer may also undergo structural changes. This was investigated by performing various heat treatments on oxide test samples. The etch rate is used as an indication of material density since high density corresponds to low etch rates. Figure 3 shows that the etch rate falls sharply with anneal temperature until approximately 800 °C where the rate of change reduces. The oxide thickness is linearly dependent to the temperature with 5% reduction observed at

1000 °C as shown in figure 4. This shrinkage in the oxide provides addition space for the molten germanium to flow and may contribute to the crack formation.

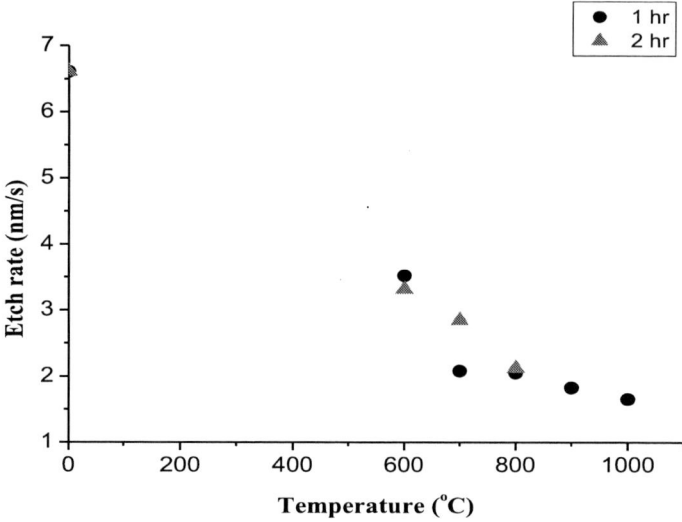

Figure 3. Graph of oxide etch rate (nm/s) versus annealing temperature (°C).

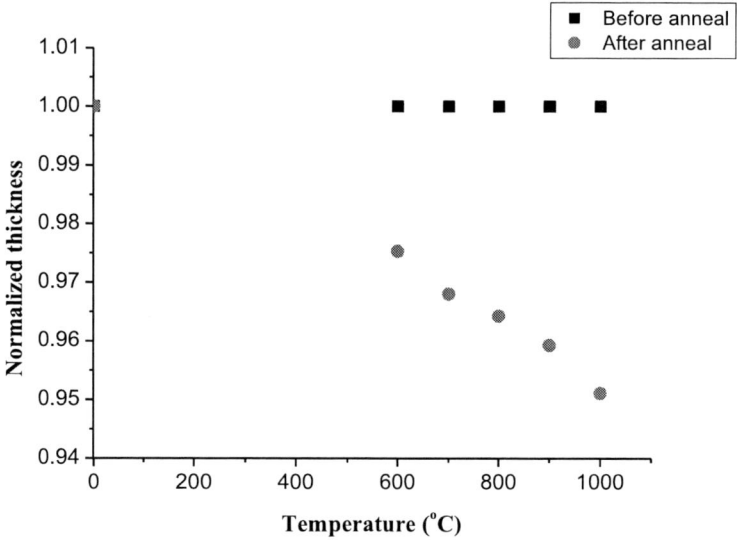

Figure 4. Normalized graph of oxide thickness versus temperature (°C) before and after the annealing process.

The micro-crucible was reinforced by adding a 1μm thick poly-crystalline silicon (polysilicon) layer on top of the PECVD oxide capping layer. The polysilicon was deposited by low pressure chemical vapor deposition (LPCVD) at 620 °C. Addition of this layer was found to minimize germanium balling and prevent delamination, with a smooth layer of germanium stripes observed in figure 5(a) and (b). These defect free structures are observed for 60 to 400μm length with 170nm thick PVD germanium stripes. A 1 second RTA at 942 °C was used.

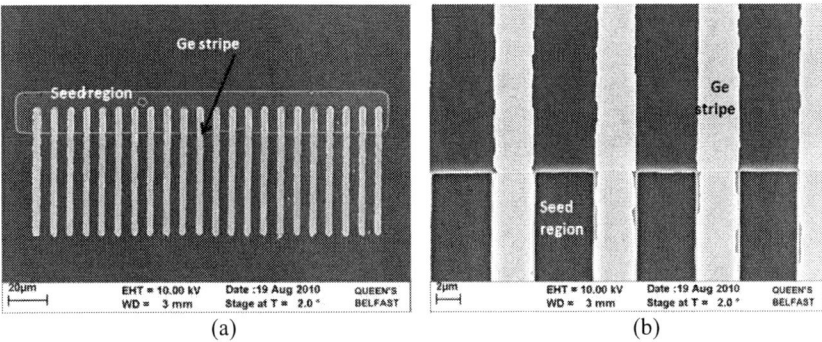

(a)   (b)

Figure 5. (a) and (b) Micrographs showing germanium stripes produced using micro-crucible comprising polysilicon on top of PECVD oxide.

The crystalline structure of 170nm thick PVD germanium stripe was analyzed by micro-Raman Spectroscopy with bulk (single crystal) germanium used as a reference. The Raman spectrum of bulk crystalline Ge demonstrates an intense longitudinal-optical phonon (LO) Ge-Ge mode with peak position at 300.2 cm^{-1} and the linewidth (or full width at half maximum (FWHM)) of ~ 3 cm^{-1}. The presence of tensile or compressive strain in the structure will result in shift of this LO peak to the low- or high-frequency side, respectively. Prior to rapid melt growth (RMG) the Raman spectrum in figure 6 demonstrates a relatively wide Ge-Ge peak with a FWHM of ~8 cm^{-1} compared to the bulk germanium reference sample with FWHM = 3.2 cm^{-1}. Along with the asymmetry, observed from the low frequency side of the peak, this suggests the presence of some amorphous content as well as polycrystalline nature of the structure. This partial crystal growth occurs in the germanium during the deposition of the capping layer at up to 620 °C. After RMG the Raman spectra (figure 6) shows a sharp and symmetric peak (with FWHM of <4 cm^{-1}) indicating an improved crystalline quality of Ge stripe compared to the unannealed sample.

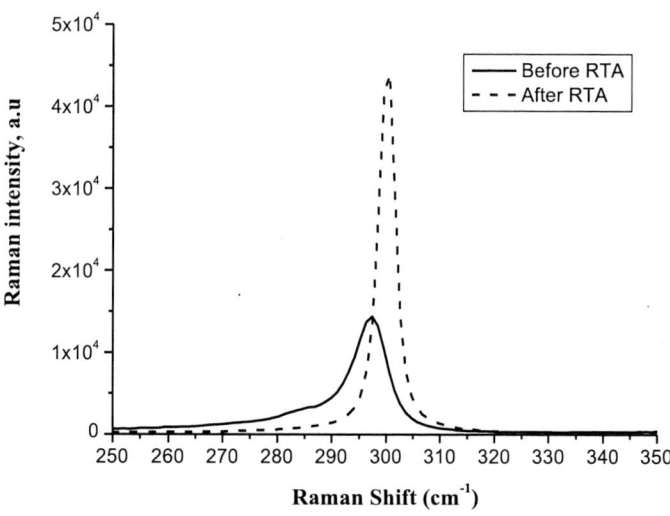

Figure 6. Raman spectra before and after rapid thermal anneal.

The comparison of Raman spectra, registered from the Ge stripe annealed at 942 °C after RMG and from n-type bulk Ge (reference sample) is shown in figure 7. One can see that the spectra are practically identical, indicating good crystalline quality germanium in the stripe.

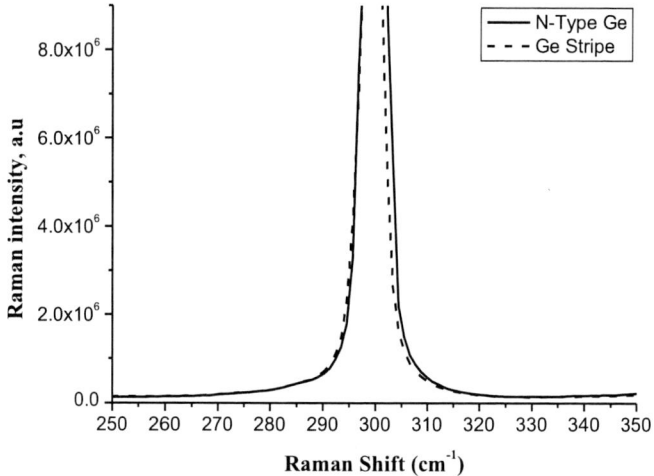

Figure 7. Spectral comparisons of germanium stripe after RMG process with single crystal germanium.

Micro-Raman line-mapping measurements were also carried out along the length of PVD germanium stripe for samples annealed at higher temperature (980 °C) with thickness of 170 nm and 210 nm. A uniform stress distribution along the stripe is obtained for both thicknesses with Ge-Ge peak position at 299.5 cm^{-1} as shown in figure 8, slightly lower than that for the bulk germanium at 300.2 cm^{-1}.This shift in peak position indicates a small tensile stress which is due to the difference of thermal expansion coefficients between the germanium and micro-crucible materials. A discontinuity in the Raman line-mapping profiles is observed at the edge of the seed window. This is thought to be due to a change in surface gradient and in some cases the presence of micro cracks after removal of the capping layer.

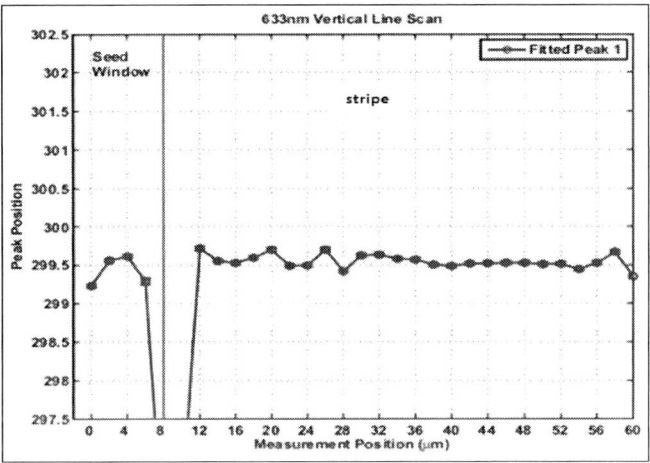

Figure 8. Peak position of germanium finger stripe annealed at 980 °C showing a uniform stress distribution along the stripe outside the seed window.

The crystalline quality of the sample annealed at 980 °C is also improved compared to the sample treated at 942 °C with the FWHM for the 170nm germanium stripe of ~3.3 cm^{-1} [9] being almost identical to that for the bulk Ge (~3.2 cm^{-1}). A slightly increased linewidth (FWHM ~3.6 cm^{-1}) is observed for the 210nm thick germanium stripe due to the rougher surface. Figure 9 shows that the spectrum from the seed area has a wider peak compared to the stripe region due to dislocations caused by silicon and germanium lattice mismatch. The dislocations do not extend along on the germanium stripe.

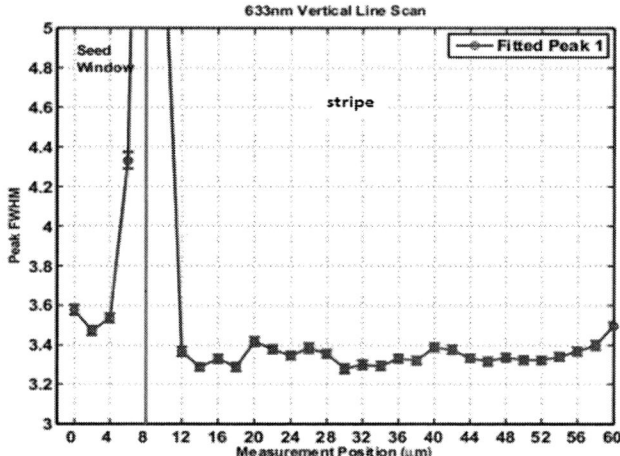

Figure 9. FWHM peak of germanium stripe annealed at 980 °C showing dislocation does not extend along the stripe.

In case of the CVD samples, no cracks and delamination is observed after annealing at higher temperature (≥ 1000 °C) and removal of capping layer. However, a rougher surface is produced with less uniform stress distribution revealed by the Ge-Ge peak position and a degraded germanium crystalline quality as confirmed by larger linewidth (with FWHM varied within ~5–8 cm^{-1}) obtained during line mapping Raman measurements. In contrast, PVD films show a longer distance of uniform stress distribution and FWHM as annealing temperature increases.

Insulator Layer and Substrates. During the rapid melt process, the molten germanium is in contact with the inner walls of the micro-crucible. Interaction between the germanium and these surfaces can influence the structure of the resultant germanium stripes. Typical dielectric materials used for the crucible are silicon dioxide or silicon nitride. Hafnium dioxide is an alternative material which has a high dielectric constant and is employed in the gate stack of advanced MOS devices. Early work has shown that the use of silicon nitride does not allow good quality crystallized germanium films to be obtained due to a significant tendency for the molten germanium to balling.

Hafnium dioxide (HfO$_2$) is investigated as insulator and crucible layer and compared to the silicon dioxide. A 20nm HfO$_2$ layer is deposited on top of oxidized silicon using atomic layer deposition (ALD) and patterned to form a seed window. A 170nm thick PVD germanium layer is then deposited and patterned into narrow stripes which run from seed window along the hafnium dioxide surfaces. These stripes are then covered with a 20nm thick hafnium dioxide, 1μm of PECVD oxide and 1μm of polysilicon that act as capping layers. A higher annealing temperature (≥ 1000 °C) has been applied. The polysilicon and PECVD oxide capping has been removed using polysilicon etch and buffered HF after the annealing process. However, HfO_2 cap is difficult to etch after high temperature treatment, therefore the HfO_2 cap layer is not removed.

In case of SiO_2 as insulator layer, cracks in the stripe occur at the edge of the seed window after removal of the polysilicon and oxide cap for the170nm germanium thickness after the RMG process. However, the stripe has a smooth surface as shown in figure 10 (a). No cracks are observed at the step between seed HfO_2 window and stripes for 60μm to 400μm. On the other hand, the germanium surface is rougher as observed on the SEM images shown in figure 10(b).

 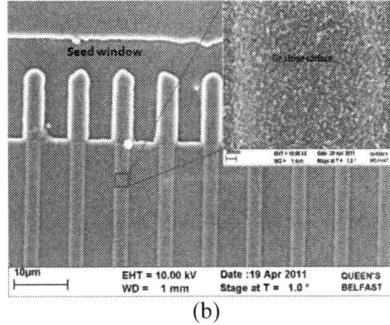

(a)   (b)

Figure 10. Micrograph of (a) germanium stripe on silicon dioxide after annealing at 980 °C and (b) HfO_2 samples after $\geq$ 1000 °C annealing temperature for 60μm distance.

Compared to germanium on silicon dioxide the FWHM of LO Raman peak registered from Ge stripe on hafnium dioxide is slightly wider ($\sim$3.9 cm^{-1}), this is believed to be due to the increased surface roughness. The peak position observed at 299.3 cm^{-1} is slightly lower than that for the bulk Ge at 300.2 cm^{-1} as expected, indicating a small tensile stress due the differences thermal expansion coefficient of materials. In terms of semiconductor devices, tensile strain in germanium is useful as it provides a significant impact on the band structure that will results in increase of the electron mobility and thus, enhance the optoelectronics properties [10, 11].

Silicon Germanium (SiGe) Content. In case of CVD germanium, it is necessary to grow first a thin silicon layer on the oxide to enable the selective germanium deposition. This is because it is impossible to deposit germanium by CVD on the oxide but only selective onto silicon. Therefore, thin silicon layer is required to be deposited on oxide prior to the CVD germanium deposition. After the RMG process, the Raman spectrum shows a sharp Ge-Ge peak at 300 cm^{-1} with FWHM of < 4 cm^{-1} for both CVD and PVD samples. This indicates a good crystalline quality of germanium stripes comparable to the bulk Ge (FWHM = 3.2 cm^{-1}) as shown in figure 11. The peak position for both samples is slightly shifted to the low-frequency side due to the tensile strain. A peak corresponding to SiGe bonding is observed in the Raman spectrum around 380-390 cm^{-1} in case of CVD sample annealed at $T \geq$940 °C.

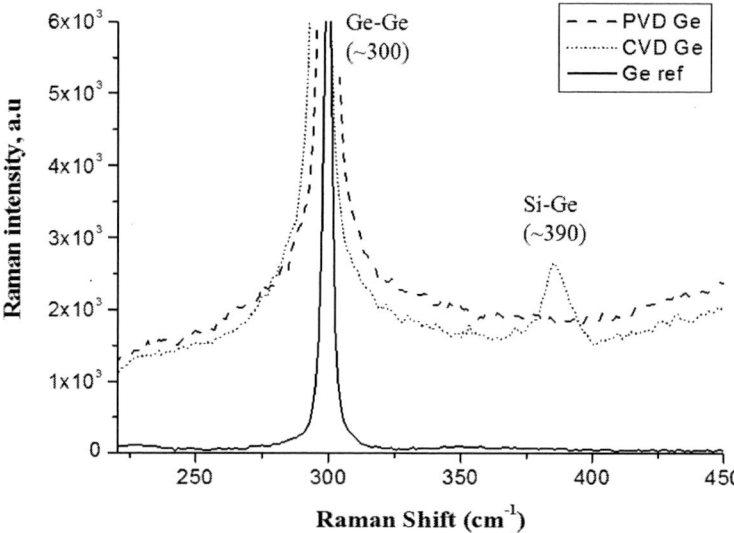

Figure 11. Raman spectra for bulk germanium reference, PVD and CVD samples after RMG at a distance of 36μm from the seed window.

The presence of SiGe is believed to result from the thin silicon layer. Miyao et al. [7] observed a similar SiGe peak in the RMG germanium grown by molecular beam epitaxy and attributed this to silicon diffusion from the seed window. In our PVD samples SiGe was not found even for high temperature (≥1000 °C) annealing where diffusion of silicon from the seed window might be expected.

It has been suggested that the lateral crystallization is driven by either silicon concentration gradient [7] or thermal gradient from the seed window [12] or a combination of both [12, 13]. In the case of the thermal gradient, the seed window provides a heat sink during temperature ramp down at the end of the RTA. This causes the germanium close to the seed to solidify first and crystallization progresses along the stripe. In the silicon gradient case, the higher silicon content at the seed results in a higher melting point and again on cooling this region solidifies first. To explore the driving mechanism the micro-crucible was fabricated on a sapphire substrate, producing a structure with no source of silicon.

A 50nm silicon dioxide layer was deposited on the sapphire using PECVD and patterned to make a seed window. A 170nm thick PVD germanium layer is deposited covering the seed area and silicon dioxide and then patterned into narrow stripes. A 1μm PECVD oxide and 1μm polysilicon layers were deposited on germanium to act as the capping layer. After RTA at 985 °C, the capping layer was removed and Raman spectroscopy was used to investigate the crystalline quality. As expected no SiGe or silicon peaks were present. The Ge-Ge peak shown in figure 12 has a FWHM value of

~6.4 cm^{-1} as well as asymmetry from low-frequency side of spectra, indicating that the Ge stripe has polycrystalline structure with some amorphous content. This suggests that the presence of silicon is necessary for optimum crystalline quality. No delamination is observed on the SEM image with 170 nm thick PVD germanium. It was also found that the sapphire substrate provides less tensile stress (as negligible shift in peak position is observed). This is due to the thermal expansion coefficient of sapphire being more closely matched to that germanium compared to silicon.

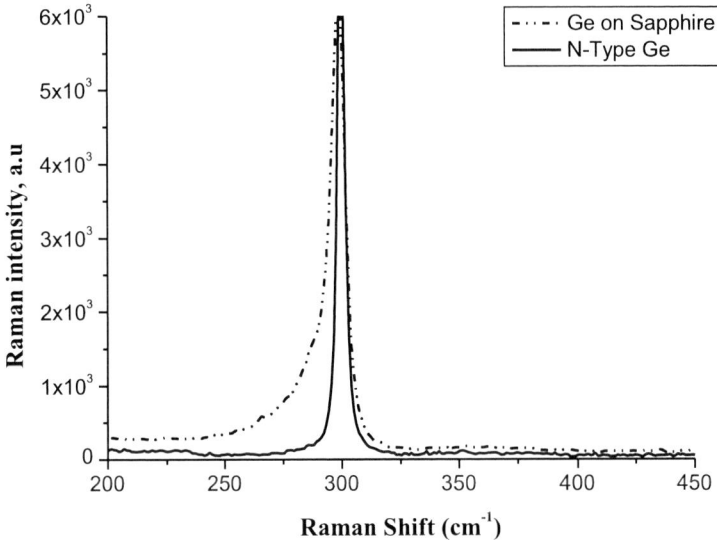

Figure 12. Raman spectra of germanium on sapphire substrate.

Conclusions. PECVD oxide capping alone is not rigid enough to hold the molten germanium resulting in cracks and delamination. Reinforcing the crucible with a polysilicon cap on top of the PECVD oxide provides robust capping layers thus prevents balling and delamination. In contrast to thermal treatment at 942 °C, a higher annealing temperature provides a better crystalline quality of PVD deposited Ge films. This is confirmed by uniformity of Raman spectra characteristics (the peak position and FWHM) extended further away from the seed window. The rougher surface of PVD Ge films on hafnium dioxide results in a slight widening of the Ge spectra. In contrast to samples with CVD germanium, no SiGe spectrum was traced for PVD samples even annealing at high temperature (≥1000 °C). This proves that the intermixing of Si and Ge is due to the thin silicon layer deposited prior to the germanium in the CVD process and not because of the diffusion of silicon from the seed window. CVD samples demonstrate less uniform Raman spectra with degradation of germanium crystal quality as annealing at high temperature (≥1000 °C) in contrast to PVD films. In the absence of a silicon seed, which was the case for samples on sapphire substrate, a lower quality germanium crystal structure (polycrystalline with some amorphous content) was observed. This indicates that silicon diffusion is an important mechanism in the rapid melt process. It is believed that the thermal flow from seed window is causing the germanium lateral growth. Results from micro-Raman Spectroscopy shows that good quality of germanium stripes can be obtained from rapid melt growth (RMG) process. This technology thus has potential to produce a low cost, high-quality future semiconductor solar cell.

Acknowledgments

The authors wish to thank the financial support from Ministry of Higher Education Malaysia and the Royal Society/Royal Irish Academy. D.A. acknowledges the financial support from IRCSET (Ireland) Postgraduate Award.

References

1. S. S. Sun, N. S. Sariciftci, *Organic Photovoltaics Mechanism, Materials and Devices*, Editors, p. 19, Taylor & Francis Group (2005).
2. A. Ennaoui, S. Fiechter, Ch. Pattenkofer, N. Alonso-Vante, K. Buker, M. Bronold, Ch. Hopfner and H. Tributsch, *Sol. Energy Mater. Sol. Cells* **29**, (1993).
3. G. W. Crabtree and N. S. Lewis, *Physic today*, **60**, (2007)
4. C. Wadia, A. P. Alivisatos and D. M. Kammen, *Environ. Sci. Technol.* **43**, (2009).
5. B. Bitnar, *Semicond. Sci. Technol.*, **18**, (2003).
6. Y. Liu, M. D. Deal, J. D. Plummer, *J. Electrochem. Soc.*, **152**, 8 (2005).
7. M. Miyao, K. Toko, T. Tanaka and T. Sadoh, *App. Phys. Lett.* **95**, 022115 (2009).
8. S. Balakumar, M. M. Roy, B. Ramamurthy, C. H. Tung, G. Fei, S. Tripathy, C. Dongzhi, R. Kumar, N. Balasubramanian and D. L. Kwong, *Electrochem. Solid-State Lett.*, **9**, 5 (2006).
9. N. Zainal, S. J. N. Mitchell, D. W. McNeill, M. F. Bain, B. M. Armstrong, P. T. Baine, D. Adley and T. S. Perova, *UK Semicond. Conf.*, (2011).
10. J. Michel, J. Liu and L. C. Kimerling, *NPhoton* **4**, 157 (2010).
11. E. E. Haller, *Mat. Sci in Semicond. Pros.* **9**, (2006).
12. K. Toko, T. Sakane, T. Tanaka, T. Sadoh and M. Miyao, *App. Phys. Lett.* **95**, 112107 (2009).
13. T. Sadoh and M. Miyao, *ECS Trans.*, **33**, 6 (2010).

CHAPTER 6

FABRICATION AND CHARACTERIZATION OF III-V'S

ECS Transactions, 45 (4) 183-188 (2012)
10.1149/1.3701133 ©The Electrochemical Society

In-situ As$_2$ decapping and atomic layer deposition of Al$_2$O$_3$ on n-InGaAs(100)

Jaesoo Ahn, Byungha Shin and Paul C. McIntyre

Department of Materials Science & Engineering,
Stanford University, Stanford, CA 94305, USA

InGaAs(100) channel surfaces that were originally covered with a protective As$_2$ layer are thermally decapped in-situ in a high vacuum atomic layer deposition system prior to Al$_2$O$_3$ deposition. The As$_2$ decapping process is monitored by chamber pressure change and in-situ x-ray photoelectron spectroscopy. A pressure spike is observed during the decapping process and the completion of the As$_2$ desorption is confirmed by a chemical shift in As 3d bonding to the lower binding energy by 0.8 eV. Post-metallization forming gas anneal improves the interface properties and lowers the density of defects at the oxide/InGaAs interface by passivating interface defects and border traps. Thermal evaporation of the gate metal further reduces the frequency dispersion in weak inversion and accumulation compared to the same metals deposited by e-beam evaporation.

Introduction

Although III-V semiconductors and high-κ dielectrics are attracting great interest due to their potential application in highly scaled high-mobility field effect devices, developing a stable interface with a low density of defects between high-κ gate dielectrics and III-V channel layers remains a main challenge for III-V based metal-oxide-semiconductor field-effect transistors (MOSFETs) (1). In$_{0.53}$Ga$_{0.47}$As and atomic layer deposited (ALD) Al$_2$O$_3$ are among the leading candidates for high-κ/III-V n-channel MOS devices because of their high electron mobility and stable interface with relatively low interface defect density (2, 3). However, since III-V semiconductors are liable to be oxidized and their native oxides have poor electrical quality, suppressing oxidation of the III-V semiconductor surface prior to oxide deposition and passivating defect sites after the dielectric deposition are crucial to achieve superior transistor performance (4). Therefore, many researches have focused on III-V surface treatment (5) and passivation of III-V surface post oxide deposition (6) in order to reduce the defect density. In this article, we present the use of a protective As$_2$ capping layer on n-InGaAs surface to prevent oxidation of InGaAs surface and *in-situ* desorption of the As$_2$ layer in a high vacuum ALD chamber prior to Al$_2$O$_3$ deposition. The decapping process is monitored by pressure change and in-situ x-ray photoelectron spectroscopy (XPS). Electrical properties of the As$_2$ decapped InGaAs surface are also described.

Experiments

Epitaxial n-In$_{0.53}$Ga$_{0.47}$As(100) layers with Si doping concentration of 2×10^{16} cm^{-3} were grown on heavily doped n-type InP substrate by molecular beam epitaxy (MBE) and covered with an amorphous As_2 capping layer of 80 ~ 120 nm thickness to protect the InGaAs surface from oxidation during air exposure. The arsenic capping layer was completely removed by thermal desorption process *in-situ* in a high vacuum ALD chamber prior to oxide deposition. The decapping process was monitored by pressure change and confirmed by *in-situ* XPS measurement. Al_2O_3 depositions using TMA and H_2O pulses (TMA-first in sequence) were followed at 270°C after the decapping of the As_2 layers. 50 nm thick palladium gate electrodes were deposited by either thermal evaporation or e-beam evaporation through a shadow mask, and wafer back side contacts of 50 nm Au/20 nm Ti were used to reduce the contact resistance. Post-metallization forming gas (5% H_2/95% N_2) anneal (FGA) was done for 30 minutes at 400°C. Multi-frequency capacitance-voltage (C-V) was measured in the frequency range from 1kHz to 1MHz at room temperature in the dark, using a HP4284A precision LCR meter.

Results and Discussion

In-situ As_2 Decapping. $In_{0.53}Ga_{0.47}As(100)$ channel surface capped with an amorphous As_2 layer is baked at 150°C for at least 1 hour to remove residual moisture in the As_2 layer and then heated up slowly in the ALD reactor where the initial chamber pressure was ~ 2×10^{-7} Torr. As the substrate temperature increases at the rate of 5 ~ 6 °C/min, the chamber pressure undergoes a sharp increase and suddenly drops back to the initial chamber pressure when the arsenic layer is fully desorbed. The decapping temperature and the shape of the chamber pressure peak depend on the thickness of the arsenic capping layer. Figure 1 shows the substrate temperature and the ALD chamber pressure obtained during the decapping process with a thin and a thick arsenic layer. As shown in Fig. 1(a), when the arsenic layer is thin (~80nm or thinner), the decapping process is completed at the substrate temperature of 360~370°C and very sharp pressure peak is observed. On the other hand, if a thicker arsenic layer is deposited on InGaAs substrate, the arsenic layer is desorbed at higher temperature (380~390°C) and a broader pressure peak is observed.

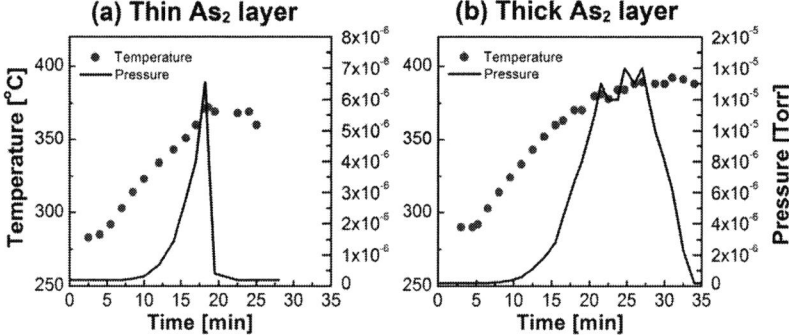

Figure 1. Arsenic decapping process represented by substrate temperature and ALD reactor pressure when the arsenic layer is (a) thin and (b) thick. A pressure spike is observed.

Because the arsenic layer is deposited during cool down from the growth regime of the InGaAs layer in a MBE chamber, the density of the arsenic layer at the initial stage of As_2 deposition may be different from the density at the late stage of As_2 deposition. Therefore, denser region of the As_2 layer may be decapped at higher temperature.

The completion of the arsenic decapping is confirmed by the chemical binding energy shift of the As_2 layer from the *in-situ* XPS measuremt. Since the *in-situ* XPS spectrometer is directly equipped on the ALD chamber, neither sample transfer to another measurement chamber nor change in substrate temperature is needed for the measurement after a deposition step (7). In order to minimize the attenuation of photoelectrons in the high process pressure of the ALD chamber and to collect as many photoelectrons as possible from the samples into the energy analyzer, the x-ray source and the electron lens are differentially pumped and the aperture of the electron lens is very closely placed from the sample surface. Figure 2 is a schematic diagram of the *in-situ* XPS system. A detailed description of this system and of ALD combined with *in-situ* XPS can be found in Ref. 7 and Ref. 8.

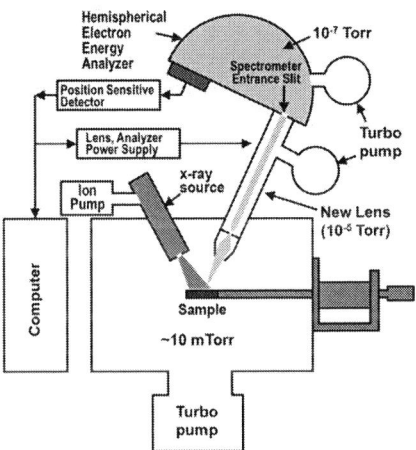

Figure 2. A schematic diagram of the differentially pumped *in-situ* XPS spectrometer using Al K_α source with a characteristic energy of 1486.6 eV.

In-situ XPS spectra show that only arsenic peaks are present in the As_2-capped InGaAs surface and no substrate peak is observed because the As_2 layer is thicker than the depth resolution of the x-ray source. After As_2 decapping, the substrate peaks (In and Ga peaks) appear (not shown here) and a chemical shift in As 3d peaks is observed. Figure 3 shows the *in-situ* XPS As 3d spectra taken before and after As_2 decapping. The initial binding energy of As 3d core level is 41.8 eV, which is consistent with As-As metallic bonding. After the decapping process, the As 3d peak shifts to As-In and As-Ga bonding (41.0 eV) as shown in Fig. 3, indicating the complete removal of the As_2 layer.

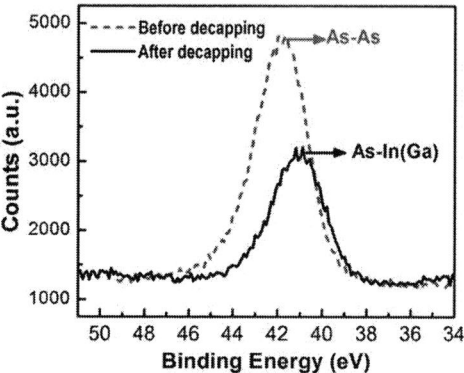

Figure 3. XPS As 3d core level spectra before and after As_2 decapping. Chemical shift to the lower binding energy by ~0.8 eV is observed.

ALD Al_2O_3 and Metallization After the arsenic layer is desorbed, MOS capacitors with ~2.5 nm Al_2O_3 are fabricated for electrical measurements. Figure 4(a) illustrates multi-frequency C-V characteristics of as-deposited Al_2O_3/n-InGaAs capacitor with e-beam evaporated Pd as a gate electrode. The as-deposited Al_2O_3 exhibits large frequency dispersion throughout the applied gate bias range, which shows that the as-grown oxide contains large density of interface traps and border traps. These defects are effectively passivated and/or removed by post-metallization FGA at 400°C for 30 min. As a result, the frequency dispersion is reduced as shown in Fig 4(b). In addition, the flatband voltage is shifted close to the ideal value (~0.5 eV, given by the work function difference between gate metal and semiconductor), which is indicative of reduction of the oxide trapped charges.

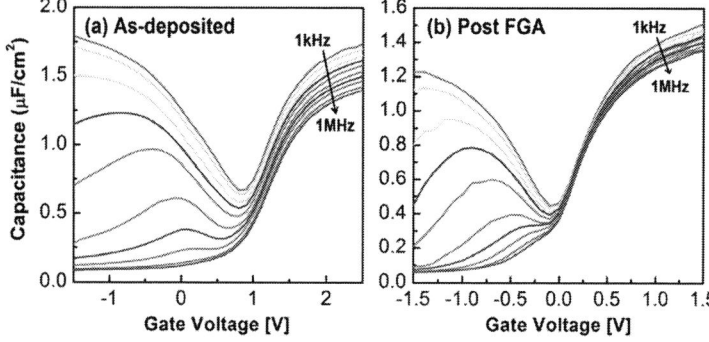

Figure 4. Multi-frequency (1kHz to 1MHz) C-V characteristics from (a) as-deposited and (b) forming gas annealed Al_2O_3/n-InGaAs with e-beam evaporated Pd as a gate electrode.

Some amount of interface defects and border traps can be created by gate metal deposition after Al_2O_3 atomic layer deposition. Figure 5(a) shows the post-FGA C-V characteristics of the capacitor with thermally evaporated Pd as a gate electrode. The multi-frequency C-V curves clearly show reduced frequency dispersion in weak inversion and accumulation. Figure 5(b) shows the frequency dispersion measured at 1.5 V and -1.0 V (inset of Fig 5(b)). The frequency dispersion in accumulation is reduced from 3.38 %/decade to 2.67 %/decade by using the thermal evaporation process. It has been reported that e-beam evaporation process generates soft x-ray, which can cause a significant degradation of oxide/III-V semiconductor interface properties (9). Therefore, by choosing a benign process after oxide deposition, the number of interface defects and border traps can be significantly reduced.

Conclusion

In conclusion, the details of *in-situ* As_2 decapping process have been reported. The As_2 capping layer can be thermally desorbed at the substrate temperature of 360°C~390°C depending on the As_2 layer thickness. The binding energy shift of As 3d peak is observed in the differentially pumped *in-situ* XPS measurement after the complete removal of As_2 layer. Although the ALD-Al_2O_3/n-InGaAs MOSCAP shows large density of interface defects, substantial amount of the interface defects can be passivated by post-metallization FGA. Moreover, the gate electrode deposited by thermal evaporation process appears to be effective in reducing the interface defect density.

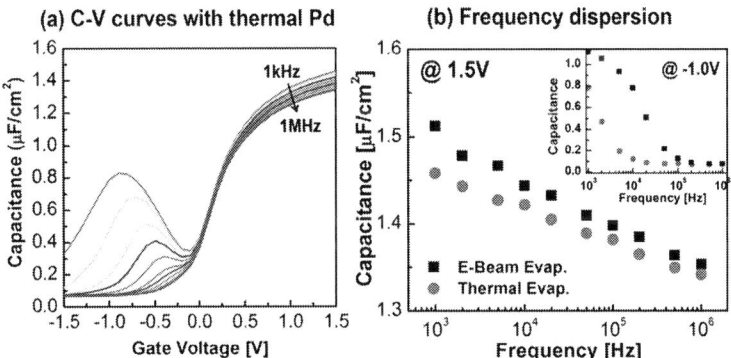

Figure 5. (a) Multi-frequency (1kHz to 1MHz) C-V curves from Al_2O_3/n-InGaAs with thermally evaporated Pd as a gate electrode. (b) Comparison of frequency dispersion in accumulation measured at 1.5 V between the capacitors with e-beam evaporated Pd and thermally evaporated Pd. The inset shows the frequency dispersion in inversion measured at -1.0 V

Acknowledgments

The authors acknowledge support from Semiconductor Research Corporation through the Non-Classical CMOS Research Center (Task ID 1437.008) and the Stanford Initiative in Nanoscale Materials and Processes (INMP). We thank M. Holland and I. Thayne of University of Glasgow for supplying the As_2-capped InGaAs/InP substrates used in this work.

References

1. R. M. Wallace, P. C. McIntyre, J. Kim, and Y. Nishi, *MRS Bull.*, **34**, 493 (2009)
2. U. Singisetti, M. A. Wistey, G. J. Burek, E. Arkun, A. K. Baraskar, Y. Sun, E. W. Kiewra, B. J. Thibeault, A. C. Gossard, C. J. Palmstrom, and M. J. W. Rodwell, *Phys. Status Solidi C,* **6**, 1394 (2009).
3. B. Shin, J. B. Clemens, M. A. Kelly, A. C. Kummel, and P. C. McIntyre, *Appl. Phys, Lett.*, **96**, 252907 (2010)
4. P. C. McIntyre, Y. Oshima, E. Kim, and K. C. Saraswat, *Microelectron. Eng.*, **86**, 1536 (2009)
5. É. O'Connor, S. Monaghan, K. Cherkaoui, I. M. Povey, and P. K. Hurley, *Appl. Phys. Lett.*, **99**, 212901 (2011)
6. E. J. Kim, L. Q. Wang, P. M. Asbeck, K. C. Saraswat, and P. C. McIntyre, *Appl. Phys. Lett.*, **96**, 012906 (2010)
7. M. A. Kelly, M. L. Shek, P. Pianetta, T. M. Gur, and M. R. Beasley, *J. Vac. Sci. Technol. A*, **19**, 2127 (2001)
8. S. Swaminathan and P. C. McIntyre, *ECS Trans.*, **33**, 455 (2010)
9. G. J. Burek, Y. Hwang, A. D. Carter, V. Chobpattana, J. J. M. Law, W. J. Mitchel, B. Thibeault, S. Stemmer, M. J. W. Rodwell, *J. Vac. Sci. Technol. B*, **29**, 040603 (2011)

ECS Transactions, 45 (4) 189-201 (2012)
10.1149/1.3700468 ©The Electrochemical Society

Germanium doping, contacts, and thin-body structures

R. Duffy, M. Shayesteh

Tyndall National Institute, University College Cork, Lee Maltings, Cork, Ireland

Interest in Ge has risen sharply since the turn of the century, as it has enabled applications ranging from field-effect-transistors, solar cells, photonics, and sensors. In this paper the state-of-the-art will be reviewed in the areas of (i) dopant behavior and ultrashallow junction performance metrics, (ii) thin-body structure formation, and (iii) contact methodologies in Ge.

Introduction

Moore's law is facing extinction as fundamental roadblocks barricade the traditional miniaturization path of Si devices. The likely resolution involves alternative materials and alternative device architectures, such as the fin-based tri-gate transistor Intel recently announced in their latest technology (1). While Si has dominated the semiconductor industry for many decades, alternative materials such as Ge, III-Vs, and graphene are rapidly emerging as realistic competitors and potential novel "building blocks" for revolutionary semiconductor technologies. Ge in particular, with high carrier mobilities and narrow bandgap, has been seen in information and communications technology and energy efficient technologies applications such as field-effect-transistors (2,3), light-emission and photonic applications (4,5), image-sensors and photodetectors (6,7), nanocrystals (8), nanowires (9-11), and solar cells (12,13). Ge has the added advantage of being Si compatible, thus it can be processed side-by-side on existing Si platforms. Advances in process integration of Ge devices have proceeded rapidly over the last few years. P-type devices were demonstrated initially (14-16), as they are easier to realize than n-type due to fundamental material issues.

Key challenges we will discuss in this paper are in the areas of Ge dopant activation, annealing for parasitic resistance reduction and crystal defect annihilation, and low-resistive contact formation.

Dopants

High-concentration doping of Ge can be quite a substantial problem, as it is difficult to activate impurity atoms to a high enough level, prevent them escaping during thermal treatments, while maintaining good crystalline integrity of the semiconductor substrate. With respect to impurity doping of Ge n-type devices, the biggest challenge lies in the source and drain regions. P and As are relatively difficult to activate and diffuse quickly (17,18), leading to high access resistances and limited capability to reduce the device dimensions, respectively. Sb is a heavy ion, and during ion implantation it can cause severe substrate deformation. Conversely, the low diffusivity and high electrical activation of B and Ga in Ge makes them ideal p-type candidates for scaled junction formation (19). Figure 1 summarizes the main characteristics of n-type and p-type dopants in Ge.

189

n-type dopant behavior

P diffusion in Ge is faster at high concentrations than at low concentrations (20). Much of the same can be said of As diffusion in Ge. This is occasionally referred to as concentration-enhanced diffusion or a $(n/n_i)^2$ dependency (17). The resulting doping profile after a rapid thermal anneal is typically box-like in nature, with a flat plateau and a sharp drop-off in concentration in the tail. An easy way to reduce diffusion is to simply reduce the implant dose, but this of course will come with a sheet resistance penalty. P and As diffusion in Ge are considered to be vacancy-mediated (21-23), and control of the vacancy population via point defect engineering may reduce dopant diffusion. This fast diffusion characteristic may indeed be prohibitive for scaled technologies, however for technologies that require deep in-diffused profiles, or dopant in-diffusion from surfaces, rapid diffusion could be an advantage. The maximum activation of n-type dopants has been reported in the 5×10^{19} - 1×10^{20} cm^{-3} range (24-26). In terms of clustering behavior, fundamental modelling and theory predicts that donors deactivate via vacancies (27,28).

p-type dopant behavior

In contrast, the behavior of p-type dopants in Ge is quite straight-forward. B and Ga implants activate to very high levels while being essentially diffusion-free. The profiles exhibit a Gaussian shape in general. B implanted into preamorphized Ge produced a maximum activation = 5×10^{20} cm^{-3} after 360 °C solid-phase-epitaxy (29). Ga maximum activation = 6.6×10^{20} cm^{-3} was reported after 450 °C 1 hr anneal (30). No diffusion was observed after a 650 °C 1 hr anneal. Panciera et al. reported B sheet resistance (Rs) evolution versus anneal time and temperature (31), and changes in Rs were correlated with end-of-range defect dissolution and subsequent release of Ge interstitials. Ab initio modelling of B in Ge predicts diffusion via self-interstitials (32).

	n-type dopants (P, As, Sb)	p-type dopants (B, Ga)
Activation	Max. ~10^{20} cm^{-3}	Max. ~6×10^{20} cm^{-3}
Diffusion	Fast • Bad for scaling • Good for in-diffusion	Slow • Good for scaling • Bad for in-diffusion
Resulting profile shape	Box-like	Gaussian
Diffusion mechanism	V-mediated	I-mediated
Clustering mechanism	V-mediated	I-mediated
Best Rs vs Xj	Plasma doping	Ion implant
Best diode Ion/Ioff	Spin-on-dopant	Metal-induced-dopant-activation

Figure 1 : Summary of n-type and p-type dopant characteristics in Ge.

The Rs versus Xj performance metric

Figure 2 shows n-type dopant literature data for the Rs versus junction depth (Xj) performance metric. The contour lines are added to guide the eye, and show Rs versus Xj for constant active doping levels assuming a box-like profile. The carrier mobility for n-type Ge was taken from Fistul et al. (33). For the n-type dopants, the shallowest profiles have been produced by 5 keV As implants (24), and by plasma doping (PLAD) (34).

Figure 3 shows p-type dopant literature data for the Rs versus Xj performance metric. Again, the contour lines are added to guide the eye. The carrier mobility for p-type Ge was taken from Mirabella et al. (29). For the p-type dopants, the shallowest profiles have been produced by 2 keV B implants (35).

Anneal strategies

Metastable solubility and damage annihilation

Equilibrium solubility may not be sufficient to meet the aggressive access resistance targets at advanced device dimensions, thus above-equilibrium metastable solubility must be generated. Techniques to generate such metastable solubilities are well known in Si, and have been initially explored in Ge (36). Typically the crystalline substrate is amorphized by ion implantation, and recrystallized by thermal annealing thereafter. The formation of metastable solubility requires care during subsequent processing because further supply of thermal energy, e.g. by back-end processing, causes the metastable condition to revert back to the lower equilibrium state. Deactivation kinetics of dopants in Ge are at an early stage of exploration at this point in time.

Control of diffusion, to facilitate junction and device scaling, can be achieved by reducing the thermal budget of the annealing process. In Si applications high-temperature millisecond anneals (laser and flash) are popular and are beginning to be applied to Ge (37-40). As the melting point of Ge is 937 °C, a low-temperature process such as solid-phase-epitaxial-regrowth appears to be another solution (41).

Note that Ge is easier to amorphize than Si (42), possibly linked to less dynamic annealing during ion implant. Amorphization in Ge occurs even during a 5×10^{13} cm^{-2} dose P implant (43). Consequently it is expected that n-type implants required for highly doped regions will amorphize the Ge substrate. The amorphous-crystalline interface is very rough in general after implants into Ge (44). Severe damage from Sb implants can lead to the formation of a sponge-like honeycomb structure at the surface (45). Full-melt laser can cure this honeycomb damage structure (39).

Substrate desorption

A concern during thermal annealing of Ge substrates is desorption of Ge from the surface, which can lead to surface roughening and dopant loss. Ionnaou et al. experimentally investigated substrate loss of uncapped Ge and found it characterized by an Arrhenius equation with $E_A=2.03$ eV (46). Capping the substrate is a solution as a SiO_2 or a Si_3N_4 cap improved P dose retention. Kaiser et al. further experimentally studied this desorption effect as a function of anneal ambient (47). In conclusion, low thermal budgets, small furnace volumes, and vacuum conditions lower substrate loss.

♦ A. Satta et al., Appl. Phys. Lett. 88, 162118 (2006)   ⸻ A. Satta et al., Nuc. Instr. Meth. Phys. 257, 157 (2007)
✕ S. Heo et al., Elec. Sol. State Lett. 9, G136 (2006)   ◇ C. O. Chui et al., Appl. Phys. Lett. 87, 091909 (2005)
△ Q. Zhang et al., IEEE El. Dev. Lett. 27, 728 (2006)   ▢ D. Cammilleri et al., Thin Solid Films 517, 75 (2008)
◉ V. Mazzocchi et al., RTP Conference (2009)   ✕ C. H. Poon et al., J. Elec. Soc. 152, G895 (2005)
● R. Duffy et al., Appl. Phys. Lett., 96, 231909 (2010)   ▲ G. Hellings et al., Elec. Sol. St. Lett. 14, H39 (2011)

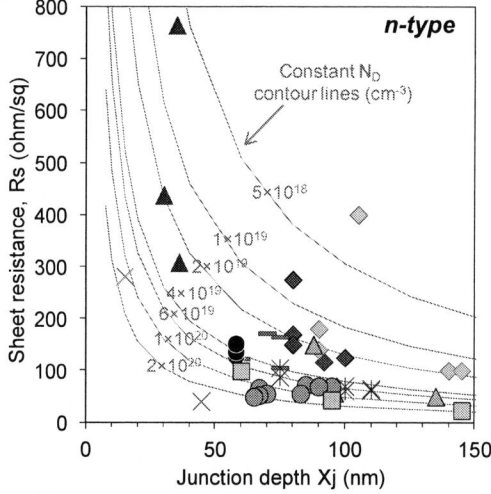

Figure 2 : Sheet resistance versus junction depth performance metric for n-type dopants in Ge. The contour lines are added to guide the eye, and show Rs versus Xj for constant active doping levels assuming a box-like profile.

♦ G. Impellizzeri et al., J. Appl. Phys. 106, 013518 (2009)
■ B. R. Yates et al., Mat. Lett. 65, 3540 (2011)   △ G. Hellings et al., Elec. Sol. State Lett. 12, H417 (2009)
✖ A. Satta et al., Appl. Phys. Lett. 87, 172109 (2005).   ◉ F. Panciera et al., Appl. Phys. Lett. 97, 012105 (2010)

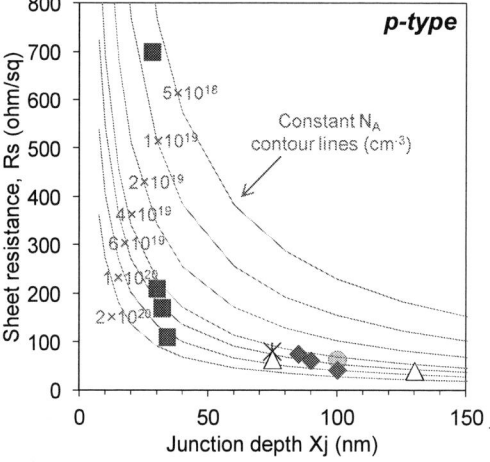

Figure 3 : Sheet resistance versus junction depth performance metric for p-type dopants in Ge. The contour lines are added to guide the eye, and show Rs versus Xj for constant active doping levels assuming a box-like profile.

Point defect engineering with co-implants

An alternative approach to control diffusion, sometimes called point-defect-engineering, is by co-implantation of a non-dopant species such as C, N or F. These species can sink point defects that cause diffusion. N implants have been demonstrated experimentally to reduce P diffusion in Ge (48,49). C also reduces P diffusion, both in co-implant (50) and molecular-beam-epitaxy form (51). A theoretical study has suggested that F should be efficient in combining with vacancy point defects and thus help to reduce vacancy-mediated diffusion (52). Recently it was shown experimentally that a F co-implant can affect the diffusion of a low dose As (3×10^{13} cm^{-2} dose) implant in certain annealing conditions (53). Another study placed F overlaying a high-concentration (1×10^{15} cm^{-2} dose) P or As dopant profile (54). It was determined that F outgasses extremely quickly from Ge, as a 1×10^{15} cm^{-2} implanted F dose escaped completely during a rapid thermal anneal as short as 1 sec, at 600 °C. This behavior is attributed to rapid diffusion, instability of F-defect clusters, and an aversion of F to reside substitutionally in the Ge lattice.

Diode behavior

Narrow bandgap

Having a bandgap of only 0.66 eV is great for photodiodes, photodetector applications, and tunnel-field-effect-transistors where enhanced carrier tunneling is generated by the narrow bandgap (55,56).

However for metal-oxide-semiconductor (MOS) applications this means a lot of unwanted off-state device leakage. To illustrate this point, Figure 4 compares experimental reverse bias diode leakage from Ge diodes (57), and Si diodes (58), as a function of substrate doping concentration. For some concentrations the Ge diode leakage is 6-7 orders of magnitude worse. It is realistic to assume that if Ge is to be used in future MOS technologies, it will be in the thin-body form where the channel can remain lowly- or un-doped, and the applications won't be at the low-standby-power end of the product chain.

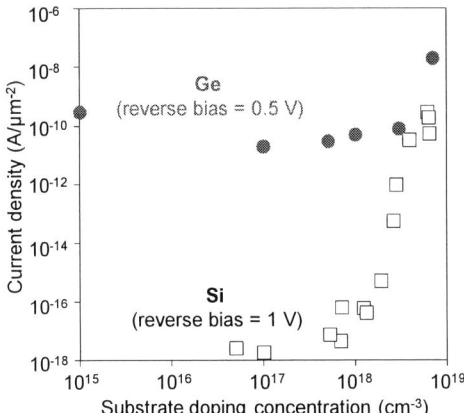

Figure 4 : Experimental reverse bias diode leakage from Ge diodes and Si diodes, versus substrate doping concentration, from (57) and (58).

The diode I_{on} / I_{off} performance metric

Diodes have been used as a diagnostic tool for several decades, and can provide information on the quality of the substrate post doping and annealing. Ideally the on current (I_{on}) should be high, while the off-current (I_{off}) should be low. In the following discussion literature data is compared for ±1 V applied in forward and reverse bias. For this performance metric I_{on} doesn't change much, while I_{off} is very sensitive to process conditions that control substrate quality and crystal defects.

Figure 5 plots I_{on}/I_{off} ratio for n+/p diodes as a function of dopant activation (or in-diffusion) anneal. Unless specified ion implant was used for the n+ doping profile. The best result came from Jamil et al. where a spin-on-dopant (SOD) approach was used to in-diffuse P (59). Other noteworthy results are the deep P implant from Chao et al. (26), and the gas-phase-doping (GPD) result from Morii et al. (60) Metal induced dopant activation (MIDA) has produced some interesting results also (61), with the added advantage of being a low-temperature process. This may be advantageous for maintaining gate stack integrity.

◆ Y.–L. Chao et al., IEEE Trans. El. Dev. 57, 665 (2010)

△ V. Ioannou-Sougleridis et al., Proc. ESSDERC, p. 367 (2009) ■ J. –H. Park et al., Appl. Phys. Lett. 93, 193507 (2008)

━ M. Jamil, IEEE El. Dev. Lett. 32, 1203 (2011) ＋ H. Shang et al., IEEE El. Dev. Lett. 25, 135 (2004)

✕ D. Kuzum et al., Proc. IEDM, p. 453 (2009) ✕ H. –Y. Yu et al., IEEE El. Dev. Lett. 30, 1002 (2009)

◉ K. Morii et al., Proc. IEDM, p. 681 (2009) ▲ S. Heo et al., Elec. Sol. State Lett. 9, G136 (2006)

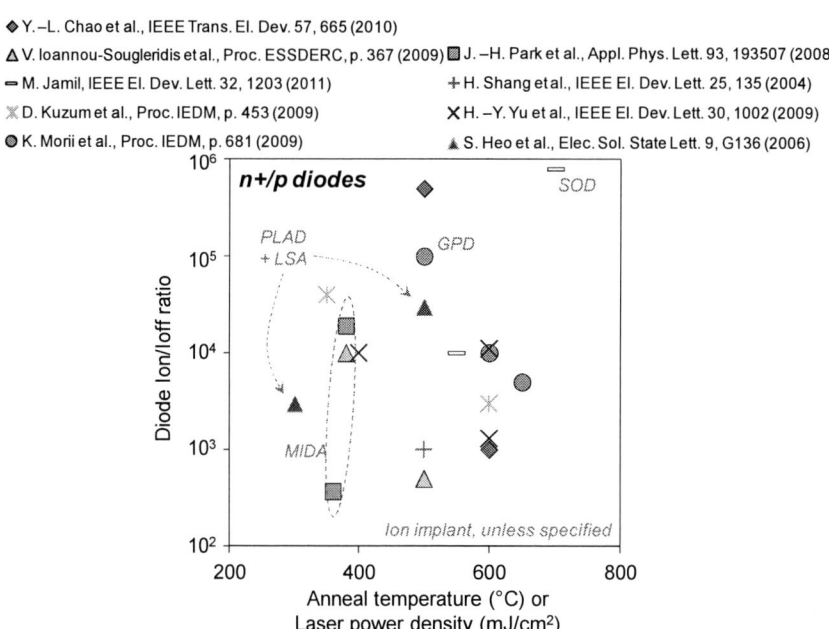

Figure 5 : Experimental data for n+/p diodes I_{on}/I_{off} ratio in Ge, for ±1V.

Figure 6 plots I_{on}/I_{off} ratio for p+/n diodes as a function of dopant activation anneal. In this case there are no in-diffusion methodologies due to p-type dopants unwillingness to diffuse in Ge. Unless specified ion implant was used for the p+ doping profile. The best result to date came from a MIDA strategy (62).

It is apparent that some of the n+/p diode results are better than the p+/n diodes. This brings us back the argument that p-type dopants lack of diffusion may be advantageous for scaling, but not good for other considerations. On this point, a large proportion of reverse bias diode leakage emanates from defects and damage in the depletion region. With a diffusing profile (n-type), the dopant can outrun this defective region, and the crystal defects can be "electrically hidden" within the high concentration profile of the junction. A non-diffusing dopant (p-type) will almost always have the end-of-range damage deeper than the implanted profile, inconveniently located in the depletion region.

■ J. –H. Park et al., Appl. Phys. Lett. 93, 193507 (2008) ◉ D. Kuzum et al., IEEE Trans. El. Dev. 56, 648 (2009)

△ V. Ioannou-Sougleridis et al., Proc. ESSDERC, p. 367,(2009) ◆ A. Satta et al., Proc. INSIGHT (2007).

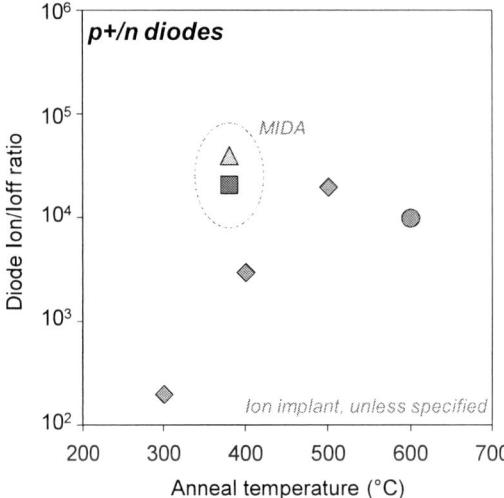

Figure 6 : Experimental data for p+/n diodes I_{on}/I_{off} ratio in Ge, for ±1V

Thin-body structures

Due to the aforementioned problem of leakage stemming from the narrow bandgap of Ge, one might assume that if Ge is used in future MOS technologies then the channel region should be undoped. The best way to implement such an architecture probably relies on either nanowire, fin-based, junctionless, or GeOI device structures.

Only a few reports exist of thin body Ge devices to date. A Ge p-channel FinFET was demonstrated by Feng et al. (63) and Ge p-type junctionless transistors were recently reported by Zhao et al. (64) with a Ge film thickness of 11.5 nm. GeOI PMOS devices were demonstrated down to L_{gate} = 30 nm (65), facilitated by thinning a Ge layer to 25 nm. As of yet, equivalent thin-body n-type devices haven't been reported.

Formation of thin-body Ge structures has been demonstrated by a number of techniques, including bonding (66), lateral-phase-epitaxy (67), aspect-ratio-trapping epitaxial growth (68), the condensation technique (69), and conventional top-down lithography (70). Ge crystal quality is the main concern for all of these approaches, as unwanted defects like twin boundaries and stacking faults are common, and must be avoided.

Contacts

Surface states associated with dangling bonds at the Ge surface can lead to Fermi-level pinning (FLP). This can seriously affect metal/Ge contact behavior as the Fermi level is pinned close to the valence band. Ohmic contacting to p-type Ge is relatively easy as a result, but the n-type system has proven difficult because of the large electron Schottky barrier height and high contact resistance (ρ_c). Figure 7 shows schematics of proposed solutions to this problem.

Ultrathin amorphous insulating layers can terminate the free dangling bonds and eliminate FLP. Another approach in creating stable low resistive contacts is to form a metal–semiconductor alloy at the surface in combination with high doping concentrations underneath.

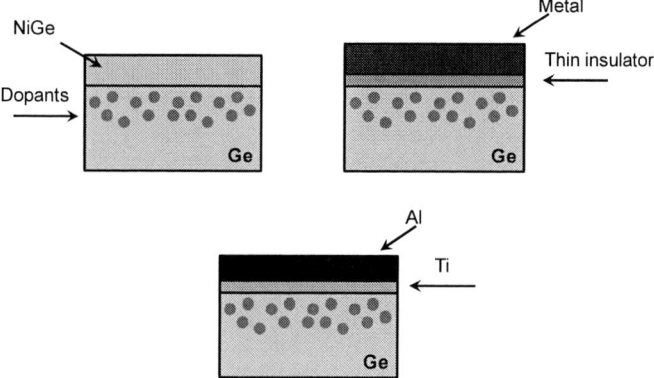

Figure 7 : Cartoon of proposed contact strategies to Ge ; (a) NiGe formation, (b) metal deposited with a thin insulator layer, and (c) a Ti contact layer in combination with Al.

Oh et al. (71) used Transfer Length Method (TLM) structures to extract ρ_c on n-type Ge, p-type Ge, n-type Si, and p-type Si, using NiSi and NiGe processes where n-type Ge was shown to be the most problematic system, producing the highest ρ_c. TLM structures were also used to extract ρ_c with Ti contacts on in-situ doped or implanted n-type Ge, as well as on SiGe (72). A record low $\rho_c = 7 \times 10^{-7}$ $\Omega.cm^2$ was reported using a process with Al/Ti contacts on Sb-doped Ge combined with laser anneal (LSA) (73). The same group recently reported $\rho_c = 8 \times 10^{-7}$ $\Omega.cm^2$ using a 500 °C anneal to activate a P/Sb co-doped n+ region (74). Alternative strategies involve NiGe combined with an As implant and LSA, or highly n-doped Si capping to reduce n-type Ge ρ_c (75). In the NiGe part of that work, ρ_c decreased with increasing LSA temperature. ρ_c was reduced by over 2 orders of magnitude switching from a 600 °C anneal to the 900 °C LSA. Figure 8 summarizes the recent work on NiGe formation on n-type Ge, as ρ_c is plotted as a function of NiGe formation temperature. The optimum temperature appears to be in the 250 – 350 °C range.

Conclusions

In conclusion, Ge process and device development may be less mature than in Si, but literature shows that rapid process has been made since the turn of the century. In this paper we reviewed the state-of-the-art in the areas of Ge dopant behavior and performance, contact formation, and thin-body structure formation. The future for Ge devices appears more optimistic as solutions are being found for problems considered to be roadblocks just a few short years ago.

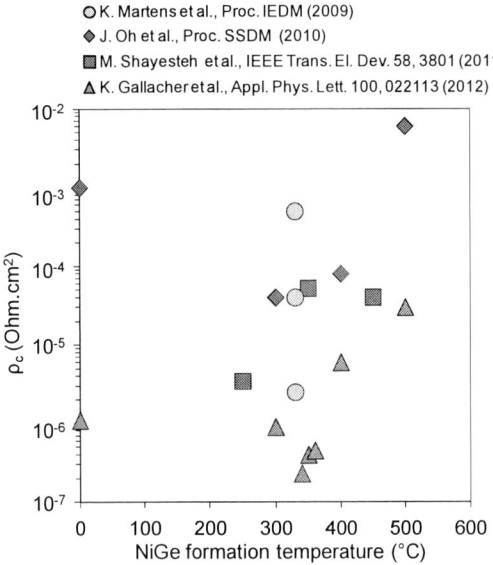

Figure 8 : Contact resistance versus NiGe formation temperature on n-type doped Ge.

Acknowledgments

This work has been funded by the Science Foundation Ireland under Research Grant No. 09/SIRG/I1623.

References

1. http://newsroom.intel.com/docs/DOC-2032
2. A. Toriumi, T. Tabata, C. H. Lee, T. Nishimura, K. Kita, and K. Nagashio, Micro. Eng. **86**, 1571 (2009).
3. F. Bellenger, B. De Jaeger, C. Merckling, M. Houssa, J. Penaud, L. Nyns, E. Vrancken, M. Caymax, M. Meuris, T. Hoffmann, K. De Meyer, and M. M. Heyns, IEEE El. Dev. Lett. **31**, 402 (2010).
4. J. Xa, K. Nemoto, Y. Ikegami, N. Usami, Y. Nakata, and Y. Shiraki, Thin Solid Films **517**, 125 (2008).
5. P. Boucaud, M. El Kurdi, S. David, X. Checoury, X. Li, T. –P. Ngo, S. Sauvage, D. Bouchier, G. Fishman, O. Kermarrec, Y. Campidelli, D. Bensahel, T. Akatsu, C. Richtarch, and B. Ghyselen, Thin Solid Films **517**, 121 (2008).
6. R. Kaufmann, G. Isella, A. Sanchez-Amores, S. Neukom, A. Neels, L. Neumann, A. Brenzikofer, A. Dommann, C. Urban, and H. von Känel, J. Appl. Phys. **110**, 023107 (2011).
7. M. Oehme, J. Werner, M. Kaschel, O. Kirfel, and E. Kasper, Thin Solid Films **517**, 137 (2008).

8. M. Klimenkov, W. Matz, S. A. Nepijko, and M. Lehmann, Nucl. Instr. and Meth. in Phys. Res. B **179**, 209 (2001).
9. S. Huang, S. –K. Shin, N. Fukata, and K. Ishibashi, J. Appl. Phys. **109**, 036101 (2011).
10. A. B. Greytak, L. J. Lauhon, M. S. Gudiksen, and C. M. Lieber, Appl. Phys. Lett. **84**, 4176 (2004).
11. S. T. Le, P. Jannaty, A. Zaslavsky, S. A. Dayeh, and S. T. Picraux, Appl. Phys. Lett. **96**, 262102 (2010).
12. D. Crisp, A. Pathare, and R. C. Ewell, Acta Astronautica. **54**, 83 (2003).
13. R. Hoheisel, J. Fernandez, F. Dimroth, A. W. Bett, IEEE Trans. El. Dev. **57**, 2190 (2010).
14. P. Zimmerman, G. Nicholas, B. De Jaeger, B. Kaczer, A. Stesmans, L. –Å. Ragnarsson, D. P. Brunco, F. E. Leys, M. Caymax, G. Winderickx, K. Opsomer, M. Meuris, and M. M. Heyns, Technical Digest of the IEEE International Electron Device Meeting 2006, San Francisco, p. 655.
15. J. Mitard, B. De Jaeger, F. E. Leys, G. Hellings, K. Martens, G. Eneman, D. P. Brunco, R. Loo, J. C. Lin, D. Shamiryan, T. Vandeweyer, G. Winderickx, E. Vrancken, C. H. Yu, K. De Meyer, M. Caymax, L. Pantisano, M. Meuris, M. M. Heyns, Technical Digest of the IEEE International Electron Device Meeting 2008, San Francisco, p. 1..
16. G. Nicholas, B. De Jaeger, D. P. Brunco, P. Zimmerman, G. Eneman, K. Martens, M. Meuris, and M. M. Heyns, IEEE Trans. El. Dev. **54**, 2503 (2007).
17. C. O. Chui, K. Gopalakrishnan, P. B. Griffin, J. D. Plummer, and K. C. Saraswat, Appl. Phys. Lett. **83**, 3275 (2003).
18. C. O. Chui, L. Kulig, J. Moran, W. Tsai, and K. C. Saraswat, Appl. Phys. Lett. **87**, 091909 (2005).
19. A. Satta, E. Simoen, T. Clarysse, T. Janssens, A. Benedetti, B. De Jaeger, M. Meuris, and W. Vandervorst, Appl. Phys. Lett. **87**, 172109 (2005).
20. S. Matsumoto and T. Niimi, J. Electrochem. Soc. **125**, 1307 (1978).
21. M. Naganawa, Y. Shimizu, M. Uematsu, K. M. Itoh, K. Sawano, Y. Shiraki, and E. E. Haller, Appl. Phys. Lett. **93**, 191905 (2008).
22. S. Koffel, P. Scheiblin, A. Claverie, and V. Mazzocchi, Mat. Sci. Eng. B **154-155**, 60 (2008).
23. S. Brotzmann and H. Bracht, J. Appl. Phys. **103**, 033508 (2008).
24. G. Hellings, E. Rosseel, E. Simoen, D. Radisic, D. H. Petersen, O. Hansen, P. F. Nielsen, G. Zschätzsch, A. Nazir, T. Clarysse, W. Vandervorst, T. Y. Hoffmann, and K. De Meyer, Elec. Sol. St. Lett. **14**, H39 (2011).
25. J. Kim, S. W. Bedell, S. L. Maurer, R. Loesing, and D. K. Sadana, Elec. Sol. State Lett. **13**, H12 (2010).
26. Y. –L. Chao and J. C. S. Woo, IEEE Trans. El. Dev. **57**, 665 (2010).
27. A. R. Peaker, V. P. Markevich, B. Hamilton, I. D. Hawkins, J. Slotte, K. Kuitunen, F. Tuomisto, A. Satta, E. Simoen, and N. V. Abrosimov, Thin Solid Films **517**, 152 (2008).
28. A. Chroneos, R. W. Grimes, H. Bracht, B. P. Uberuaga, J. Appl. Phys. **104**, 113724 (2008).
29. S. Mirabella, G. Impellizzeri, A. M. Piro, E. Bruno, and M. G. Grimaldi, Appl. Phys. Lett. **92**, 251909 (2008).
30. G. Impellizzeri, S. Mirabella, A. Irrera, M. G. Grimaldi, and E. Napolitani, J. Appl. Phys. **106**, 013518 (2009).

31. F. Panciera, P. F. Fazzini, M. Collet, J. Boucher, E. Bedel, and F. Cristiano, Appl. Phys. Lett. **97**, 012105 (2010).
32. A. Janke, R. Jones, S. Öberg, and P. R. Briddon, Phys. Rev. B. **77**, 075208 (2008).
33. V. I. Fistul, M. I. Iglitsyn, and E. M. Omel'yanovskii, Soviet Physics – Solid State **4**, 784 (1962).
34. S. Heo, S. Baek, D. Lee, M. Hasan, H. Jung, J. Lee, and H. Hwang, Elec. Sol. State Lett. **9**, G136 (2006).
35. A. R. Yates, B. L. Darby, N. G. Rudawski, K. S. Jones, D. H. Petersen, O. Hansen, R. Lin, P. F. Nielsen, and A. Kontos, Mat. Lett. **65**, 3540 (2011).
36. Y.-L. Chao, S. Prussin, J. C. S. Woo, and R. Scholz, Appl. Phys. Lett. **87**, 142102 (2005).
37. C. Wündisch, M. Posselt, B. Schmidt, V. Heera, T. Schumann, A. Mücklich, R. Grötzschel, W. Skorupa, T. Clarysse, E. Simoen, and H. Hortenbach, Appl. Phys. Lett. **95**, 252107 (2009).
38. A. Satta, A. D'Amore, E. Simoen, W. Anwand, W. Skorupa, T. Clarysse, B. van Daele, T. Janssens, Nuc. Instr. Meth. Phys. Res. **257**, 157 (2007).
39. E. Bruno, G. G. Scapellato, G. Bisognin, E. Carria, L. Romano, A. Carnera, and F. Priolo, J. Appl. Phys. **108**, 124902 (2010).
40. Q. Zhang, J. Huang, N. Wu, G. Chen, M. Hong, L. K. Bera, and C. Zhu, IEEE El. Dev. Lett. **27**, 728 (2006).
41. R. Duffy, M. Shayesteh, M. White, J. Kearney, A. –M. Kelleher, ECS Transactions, **35** (2) 185-192 (2011).
42. T. E. Haynes and O. W. Holland, Appl. Phys. Lett. **61**, 61 (1992).
43. S. Koffel, P. Scheiblin, A. Claverie, and G. Benassayag, J. Appl. Phys. **105**, 013528 (2009).
44. A. Satta, E. Simoen, T. Janssens, T. Clarysse, B. De Jaeger, A. Benedetti, I. Hoflijk, B. Brijs, M. Meuris, and W. Vandervorst, J. Electrochem. Soc. **153**, G229 (2006).
45. L. Romano, G. Impellizzeri, M. V. Tomasello, F. Giannazzo, C. Spinella, and M. G. Grimaldi, J. Appl. Phys. **107**, 084314 (2010).
46. N. Ioannou, D. Skarlatos, C. Tsamis, C. A. Krontiras, S. N. Georga, A. Christofi, and D. S. McPhail, Appl. Phys. Lett. **93**, 101910 (2008).
47. R. J. Kaiser, S. Koffel, P. Pichler, A. J. Bauer, B. Amon, L. Frey, and H. Ryssel, Micro. Eng. **88**, 499 (2011).
48. E. Simoen, A. Satta, A. D'Amore, T. Janssens, T. Clarysse, K. Martens, B. De Jaeger, A. Benedetti, I. Hoflijk, B. Brijs, M. Meuris, W. Vandervorst, Mat. Sci. Semi. Proc. **9**, 634 (2006).
49. A. Satta, E. Simoen, B. Van Daele, T. Clarysse, G. Nicholas, W. Vandervorst, W. Anwand, W. Skorupa, T. Peaker, V. Markevich, Proceedings of INSIGHT (2007).
50. V. Mazzocchi, X. Pages, M. Py, J. P. Barnes, K. Vanormelingen, L. Hutin, R. Truche, P. Vermont, M. Vinet, C. Le Royer, and K. Yckache, 17th IEEE International Conference on Advanced Thermal Processing of Semiconductors – RTP 2009, Albany (2009). D. O. I. : 10.1109/RTP.2009.5373459.
51. S. Brotzmann, H. Bracht, J. Lundsgaard Hansen, A. Nylandsted Larsen, E. Simoen, E. E. Haller, J. S. Christensen, and P. Werner, Phys. Rev. B **77**, 235207 (2008).
52. A. Chroneos, R. W. Grimes, and H. Bracht, J. Appl. Phys. **106**, 063707 (2009).
53. G. Impellizzeri, S. Boninelli, F. Priolo, E. Napolitani, C. Spinella, A. Chroneos, and H. Bracht, J. Appl. Phys. **109**, 113527 (2011).

54. M. Shayesteh, V. Djara, M. Schmidt, M. White, J. Kearney, A. –M. Kelleher, R. Duffy, Proceedings of the 19th International Conference on Ion Implantation (2012).
55. G. Han, P. Guo, Y. Yang, C. Zhan, Q. Zhou, and Y. –C. Yeo, Appl. Phys. Lett. **98**, 153502 (2011).
56. J. Zhao, A. C. Seabaugh, and T. H. Kosel, J. Elec. Soc. **154**, H536 (2007).
57. G. Eneman, M. Wiot, A. Brugère, O. S. I. Casain, S. Sonde, D. P. Brunco, B. De Jaeger, A. Satta, G. Hellings, K. De Meyer, C. Claeys, M. Meuris, M. M. Heyns, and E. Simoen, IEEE Trans. El. Dev. **55**, 2287 (2008).
58. R. Duffy, A. Heringa, V. C. Venezia, J. Loo, M. A. Verheijen, M. J. P. Hopstaken, K. van der Tak, M. de Potter, J. C. Hooker, P. Meunier-Beillard, and R. Delhougne, Solid-State Electron. **54**, 243 (2010).
59. M. Jamil, J. Mantey, E. U. Onyegam, G. D. Carpenter, E. Tutuc, and S. K. Banerjee, IEEE El. Dev. Lett. **32**, 1203 (2011).
60. K. Morii, T. Iwasaki, R. Nakane, M. Takenaka, and S. Takagi, Technical Digest of the IEEE International Electron Device Meeting 2009, Baltimore, p. 681.
61. J. –H. Park, D. Kuzum, M. Tada, and K. C. Saraswat, Appl. Phys. Lett. **93**, 193507 (2008).
62. V. Ioannou-Sougleridis, A. Dimoulas, P. Tsipas, Th. Speliotis, Proceedings of European Solid-State Device Research Conference (ESSDERC), p. 367, Athens (2009).
63. J. Feng, R. Woo, S. Chen, Y. Liu, P. B. Griffin, and J. D. Plummer, IEEE El. Dev. Lett. **28**, 637 (2007).
64. D. D. Zhao, T. Nishimura, C. H. Lee, K. Nagashio, K. Kita, and A. Toriumi, Appl. Phys. Expr. **4**, 031302 (2011).
65. L. Hutin, C. Le Royer, J. –F. Damlencourt, J. –M. Hartmann, H. Grampeix, V. Mazzocchi, C. Tabone, B. Previtali, A. Pouydebasque, M. Vinet, and O. Faynot, IEEE El. Dev. Lett. **31**, 234 (2010).
66. K.Y. Byun, P. Fleming, N. Bennett, F. Gity, P. McNally, M. Morris, I. Ferain, C. Colinge, J. Appl. Phys. **109**, 123529 (2011).
67. T. Hashimoto, C. Yoshimoto, T. Hosoi, T. Shimura, and H. Watanabe, Appl. Phys. Expr. **2**, 066502 (2009).
68. G. Wang, E. Rosseel, R. Loo, P. Favia, H. Bender, M. Caymax, M. M. Heyns, and W. Vandervorst, Appl. Phys. Lett. **96**, 111903 (2010).
69. W. van den Daele, E. Augendre, C. Le Royer, J. –F. Damlencourt, B. Grandchamp, and S. Cristoloveanu, Solid State Electronics **54**, 205 (2010).
70. M. Shayesteh, R. Duffy, B. McCarthy, A. Blake, M. White, J. Scully, R. Yu, V. Djara, M. Schmidt, N. Petkov, A. –M. Kelleher, ECS Transactions, **35** (2), pp. 27-34 (2011).
71. J. Oh, J. Huang, Y. –T. Chen, I. Ok, K. Jeon, and S. –H. Lee, Proceedings of International Conference on Solid State Devices and Materials 2010, paper P-3-20.
72. S. Raghunathan, T. Krishnamohan, and K. Saraswat, ECS Transactions, **33** (6), 871-876, (2010).
73. G. Thareja, J. Liang, S. Chopra, B. Adams, N. Patil, S. –L. Cheng, A. Nainani, E. Tasyurek, Y. Kim, S. Moffatt, R. Brennan, J. McVittie, T. Kamins, K. Saraswat, and Y. Nishi, Technical Digest of the IEEE International Electron Device Meeting 2010, paper 10.5.
74. G. Thareja, S.-L. Cheng, T. Kamins, K. Saraswat, and Y. Nishi, IEEE El. Dev. Lett. **32** (5), 608 (2011).

75. K. Martens, A. Firrinicieli, R. Rooyackers, B. Vincent, R. Loo, S. Locorotondo, E. Rosseel, T. Vandeweyer, G. Hellings, B. De Jaeger, M. Meuris, P. Favia, H. Bender, B. Douhard, J. Delmotte, W. Vandervorst, E. Simoen, G. Jurczak, D. Wouters, and J. A. Kittl, Technical Digest of the IEEE International Electron Device Meeting 2010, paper 18.4.

Germanium on Insulator (GOI) Structure Using Hetero-Epitaxial Lateral Overgrowth on Silicon

J. H. Nam, T. Fuse, Y. Nishi, and K. C. Saraswat

Department of Electrical Engineering, Stanford University, Stanford, California 94305, USA

A technique to achieve germanium on insulator (GOI) structure on Si platform using hetero-epitaxial lateral overgrowth is presented. On (100) Si wafer, SiO_2 is thermally grown, and patterned to locally reveal Si surface on which Ge is grown via selective epitaxy. After filling the growth window, Ge starts growing laterally, and adjacent Ge crystals coalesce with each other on SiO_2. After the coalescence, <100> directional growth is made dominant to fill the valley and planarize the surface. Thus, single crystalline Ge film sitting on SiO_2 can be made for GOI applications.

Introduction

Success of CMOS industry largely owes to the scalability of the Si MOSFET. As Si based devices approach their fundamental scaling limits, introducing novel materials to circumvent these limits and enhance device performance is being pursued. Higher and more symmetric carrier mobilities for electrons and holes and lower optical bandgap make Ge a promising candidate for electronics and optical devices.

Though Ge shows high bulk carrier mobility, its surface carrier mobility has been limited by poor surface passivation in the past. Also, surface Fermi level pinning near the valence band makes it difficult to get a decent ohmic contact. Recently, efforts to introduce novel gate stack and passivation layer have improved surface carrier mobilities and contact resistance, making Ge based transistors more feasible. Ge's small direct bandgap of 0.8 eV allows Ge photodetectors, optical modulators and recently light emitters to operate in the low-loss optical fiber range of about 1300 to 1550 nm and makes it a strong candidate for optical interconnect applications.

With Ge based optical and electrical devices at hand, it is needed to integrate those devices on conventional Si platform. For this, selective growth technique to grow Ge locally on Si has been studied [1], and Ge based devices on locally grown Ge on Si by using SiO_2 as growth mask have already been demonstrated [2].

Ge on insulator (GOI) based devices have several advantages. From the optical point of view, insulator layer can effectively cut off slow carrier drift–diffusion from underlying Si substrate causing bandwidth degradation, therefore gives better responsivity. On the other hand, for better electrical device performance, fully depleted Ge transistors or Ge FINFET can be fabricated on the GOI structure. Several wafer bonding processes for GOI have been studied [3], but most of them are not CMOS-compatible process, and not easy to locally integrate Ge based devices on Si based platform.

In this paper, we describe a Ge hetero-epitaxial lateral overgrowth technique coupled with selective epitaxy from a patterned substrate to get GOI structure locally grown on a Si substrate.

Process flow

300 nm ~ 2 µm thick SiO_2 film is thermally grown on a p-type (100) Si substrate at 1100 °C (Figure 1 (a)). Then, to define the area for selective Ge growth, the oxide film on the substrate is patterned using DRIE (Figure 1 (b)). After HF-last standard cleaning process, the sample is immediately loaded into an Applied Materials Centura epitaxial reactor. After high temperature hydrogen bake (1000 °C) to remove any remaining native oxide on the patterned area, thin epitaxial Si layer is grown using dichlorosilane (DCS) at 700 °C prior to the initial Ge seed layer growth at 400 °C, and then hydrogen annealing is done at 825 °C. On this seed layer, Ge is now grown epitaxially under various conditions. Initially, due to the growth selectivity between SiO_2 and Si [2], Ge first fills the growth window (Figure 1 (c)). After filling the growth window, Ge starts growing laterally, and coalesce with adjacent Ge crystals also growing laterally. After coalescence, <100> directional growth becomes dominant, and fills the valley and planarize the surface (Fig 1 (d)). Thus, single crystal Ge film sitting on SiO_2 can be realized and used for GOI applications.

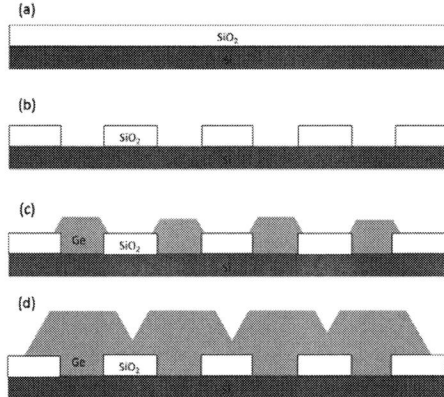

Figure 1. Schematic for Ge epitaxial lateral overgrowth process. (a) SiO_2 growth, (b) patterning, (c) Ge selective growth, and (d) lateral overgrowth

Results

Polycrystalline Ge nucleation on SiO_2 surface

Due to the limited growth selectivity, during the selective growth (Figure 1. (c)), Ge nuclei are formed on SiO_2 surface (Figure 2. (a)). Since crystalline Ge sitting on SiO_2 is to be used for GOI devices, crystalline quality of Ge in this region is crucial. To get high quality GOI, polycrystalline nucleation is suppressed by adding HCl gas during the epitaxial growth (Figure 2).

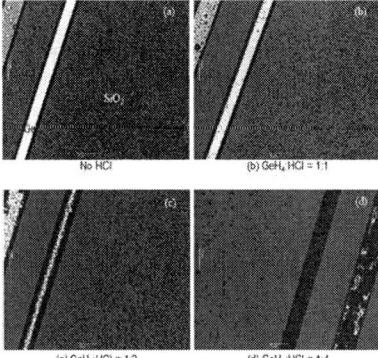

Figure 2. Suppression of Ge nucleation with different GeH$_4$:HCl flow ratio. Thin strips are Ge grown on Si, and black dots are poly Ge on SiO$_2$.

Void elimination

During the lateral overgrowth phase (Figure 1. (d)), Ge crystal shows certain crystal shape with several growth planes, and usually have negative slope at its growth front. When adjacent Ge crystals coalesce with each other on SiO$_2$ to form GOI, this negative slope leaves a void region (Figure 3). For further GOI based device fabrication, this void needs to be eliminated.

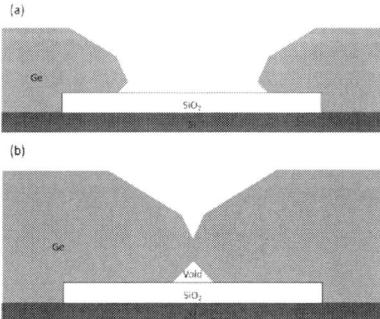

Figure 3. Schematic for void formation. Negative slope at the growth front of growing Ge crystal creates void region on SiO$_2$.

By changing the shape of the growing Ge crystal, if negative slope at the growth front can be removed, then void also can be eliminated. Shape of the Ge crystal can be engineered by adding HCl gas during the epitaxial growth, similar to the Si lateral overgrowth [4]. Figure 4. shows the influence of HCl gas on the Ge crystal shape.

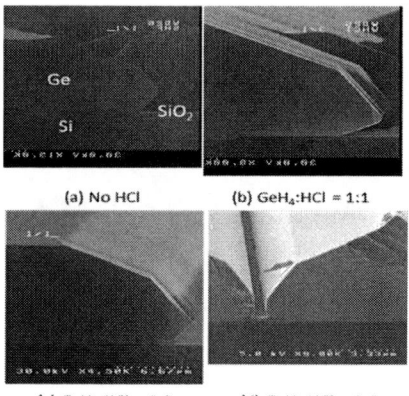

Figure 4. Facet shape with different GeH_4:HCl flow ratio during epitaxial growth. Growth at 700 °C

Pattern alignment also plays huge role in engineering the void size. If the pattern is aligned to different crystal orientations on Si substrate, then growing Ge will see different crystal planes, and resulting crystal shape is also changed. Influence of pattern orientation on void is shown in Figure 5.

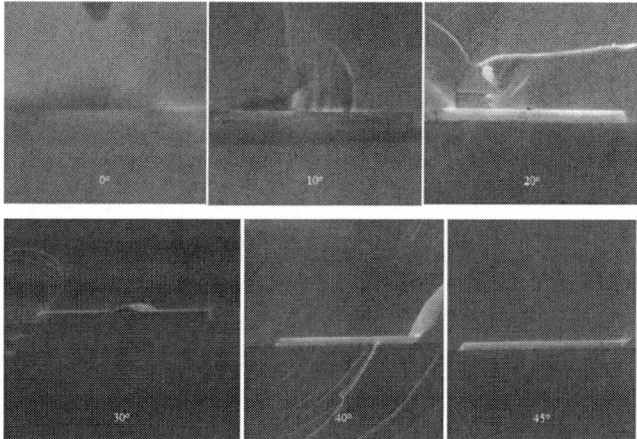

Figure 5. Effect of pattern alignment on void. 0°: pattern aligned to <110>, 45°: pattern aligned to <100>. At 30°, void size reaches its maximum, and at 40° and 45°, void is completely eliminated.

Planarization

After coalescence (Figure 1. (d)), Ge surface shows rather triangular shape with peaks and valleys (Figure 6). The surface should be planarized before fabricating any device. This can be accomplished by changing the growth condition right after the coalescence, where <100> directional growth becomes dominant. If <100> directional growth becomes dominant enough, then it will fill the valleys and make the surface planar. Figure 7 shows a void free and planarized surface of GOI film grown on Si using this technique.

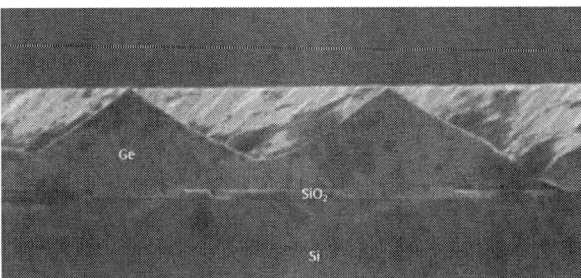

Figure 6. Ge right after coalescence. Surface needs to be planarized for further device fabrication.

Conclusion

During the epitaxial growth of Ge, HCl gas is added to suppress polycrystalline Ge nucleation on SiO_2 surface which can degrade crystal quality of Ge on SiO_2. HCl also changes the shape of the laterally growing Ge crystal, and affect the void formation. In addition to void engineering by HCl, pattern alignment is optimized to completely eliminate the void. After coalescence of neighboring Ge crystals, growth condition is changed in order to enhance <100> directional growth, which planarizes the surface. Void free crystalline GOI with planar surface is achieved, using CMOS-compatible Ge hetero-epitaxial lateral overgrowth process.

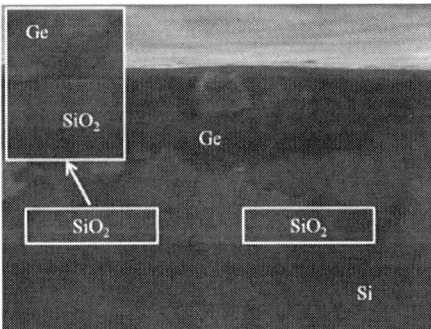

Figure 7. Void-free GOI structure using Ge selective hetero-epitaxial overgrowth.

Acknowledgments

This work was supported by U.S. Government through Advanced Photonic Integrated Circuits Corporation and the Connectivity Center of the Focus Center Research Program

References

1. J.-S. Park, J. Bai, M. Curtin, B. Adekore, M. Carroll, A. Lochtefeld, *APL*, **90**, 052113 (2007).
2. H.-Y. Yu, M. Kobayashi, W. S. Jung, A. K. Okyaya, Y. Nishi, K. C. Saraswat, *IEDM* (2009).
3. G. Taraschi, A. J. Pitera, E. A. Fitzgerald, *Solid-State Electronics*, **48**, 1297-1305 (2004).
4. D. R. Bradbury, T. I. Kamins, C.-W. Tsao, *JAP*, **55**, 519-523 (1984).

Multiple-gate In$_{0.53}$Ga$_{0.47}$As Channel n-MOSFETs with Self-Aligned Ni-InGaAs Contacts

Xingui Zhang, Hua Xin Guo, Xiao Gong, Cheng Guo, and Yee-Chia Yeo.

Department of Electrical and Computer Engineering and
Graduate School of Integrative Sciences and Engineering,
National University of Singapore (NUS), Singapore 117576.
Phone: +65-6516-2298 E-mail: yeo@ieee.org

Sub-100 nm multiple-gate In$_{0.53}$Ga$_{0.47}$As channel n-MOSFETs with self-aligned Ni-InGaAs contacts were demonstrated for the first time. With self-aligned Ni-InGaAs contacts formed on *in-situ* doped n^{++} In$_{0.53}$Ga$_{0.47}$As source and drain, the device exhibits low series resistance of 364 $\Omega \cdot \mu$m. The multiple-gate device with 50 nm channel length has a drive current of more than 411 μA/μm at $V_D = 0.7$ V and $V_G = 0.7$ V. The device also shows a peak extrinsic transconductance G_m of 590 μS/μm at $V_D = 0.5$ V.

Introduction

Much progress has been made on InGaAs channel MOSFETs to achieve high transconductance and drive current [1]-[4]. With continual scaling of device dimensions, however, the short channel effects worsen. Multiple-gate InGaAs MOSFETs have improved electrostatics for suppression of short channel effects as compared with planar InGaAs MOSFETs [5]-[9]. However, high series resistance (R_{SD}) in a multiple-gate MOSFET structure could compromise device performance. Self-aligned contacts with low contact resistance (R_C) are required to achieve low R_{SD}. Although salicide-like metallization process has been demonstrated for GaAs and InGaAs planar MOSFETs [10]-[12], there is no report of integration of salicide-like self-aligned contacts in multiple-gate InGaAs MOSFETs.

In this work, sub-100 nm multiple-gate In$_{0.53}$Ga$_{0.47}$As channel MOSFETs with self-aligned Ni-InGaAs contacts are reported for the first time. Ni-InGaAs contacts on n^{++} In$_{0.53}$Ga$_{0.47}$As S/D show very low contact resistance. With the integration of self-aligned Ni-InGaAs contacts, very low R_{SD} and good device performance were obtained. The salicide-like metallization process involves reaction of Ni with n^{++} In$_{0.53}$Ga$_{0.47}$As and a selective wet etching of unreacted Ni.

ECS Transactions, 45 (4) 209-216 (2012)

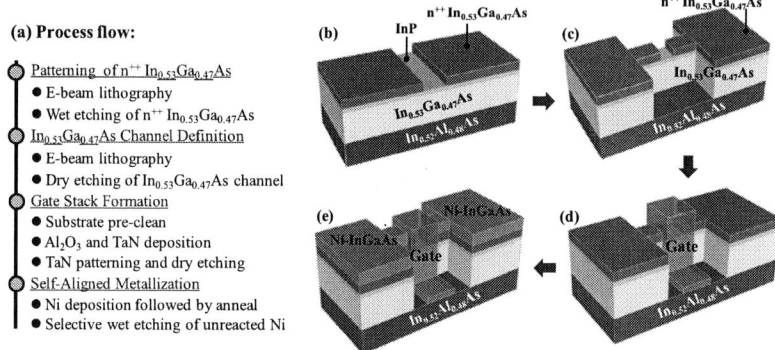

(a) Process flow:

- Patterning of n⁺⁺ $In_{0.53}Ga_{0.47}As$
  - E-beam lithography
  - Wet etching of n⁺⁺ $In_{0.53}Ga_{0.47}As$
- $In_{0.53}Ga_{0.47}As$ Channel Definition
  - E-beam lithography
  - Dry etching of $In_{0.53}Ga_{0.47}As$ channel
- Gate Stack Formation
  - Substrate pre-clean
  - Al_2O_3 and TaN deposition
  - TaN patterning and dry etching
- Self-Aligned Metallization
  - Ni deposition followed by anneal
  - Selective wet etching of unreacted Ni

Fig. 1. (a) Process flow for fabricating a multiple-gate $In_{0.53}Ga_{0.47}As$ n-MOSFET with self-aligned Ni-InGaAs contacts. (b) Schematic of the device structure after patterning n⁺⁺ $In_{0.53}Ga_{0.47}As$. (c) Device structure after definition of the $In_{0.53}Ga_{0.47}As$ channel region. (d) Device structure after gate stack formation. (e) Final device structure showing Ni-InGaAs contacts self-aligned to the gate electrode.

Device Fabrication

The fabrication process for the multiple-gate device is summarized and illustrated in Fig. 1. InP wafers served as the starting substrates. 300 nm undoped $In_{0.52}Al_{0.48}As$, 50 nm undoped $In_{0.53}Ga_{0.47}As$, 2 nm undoped InP, and 30 nm Si-doped n⁺⁺ $In_{0.53}Ga_{0.47}As$ (5×10^{19} cm⁻³) were sequentially grown. The channel region was defined [Fig. 1 (b)] by wet etching of n⁺⁺ $In_{0.53}Ga_{0.47}As$ using citric acid mixed with H_2O_2. InP served as an etch stop layer and a capping layer for the $In_{0.53}Ga_{0.47}As$ channel. Then, $In_{0.53}Ga_{0.47}As$ non-planar channel was formed by Cl_2-based plasma etching [Fig. 1 (c)]. A quick dip in H_2SO_4: H_2O_2: H_2O solution was used to smoothen the rough surface of the non-planar $In_{0.53}Ga_{0.47}As$ channel right after dry etching.

After pre-gate clean, the samples were loaded into the atomic layer deposition (ALD) tool for Al_2O_3 deposition (~6 nm). TaN metal layer was deposited by sputtering and patterned by dry etching [Fig. 1 (d)]. The devices subsequently went through the salicide-like Ni-InGaAs metallization process. 13 nm Ni was deposited by e-beam evaporator followed by a rapid thermal anneal at 250 °C for 60 s in N_2 ambient. The fabrication process was completed with selective removal of unreacted Ni in HCl solution. The final structure of the device was shown in Fig. 1 (e).

Result and Discussion

Scanning Electron Microscopy (SEM) image in Fig. 2 shows the top view of a multiple-gate $In_{0.53}Ga_{0.47}As$ channel device with Ni-InGaAs contacts self-aligned to the gate stack. The device layout is shown in the inset. The recess of n⁺⁺ $In_{0.53}Ga_{0.47}As$ defines the device channel and also separates the source and drain.

210

Transmission Electron Microscopy (TEM) image in Fig. 3 (a) shows the device cross-section along the dashed line A - A' in Fig. 2. The Ni-InGaAs contact appears as a darker layer formed on the surface of n^{++} $In_{0.53}Ga_{0.47}As$, lying adjacent and aligned to the TaN/Al_2O_3 gate stack. Ni-InGaAs has a thickness of ~22 nm with sheet resistance of about 40 Ω/square. Fig. 3 (b) shows a cross-section of the device along the dashed line B - B' (indicated in Fig. 2). The $In_{0.53}Ga_{0.47}As$ channel is in the shape of a trapezoid with a top width of 80 nm and bottom width of 170 nm. The height of $In_{0.53}Ga_{0.47}As$ channel is ~52 nm. The top surface of the $In_{0.53}Ga_{0.47}As$ channel is a (100) surface capped by 2 nm of InP while the $In_{0.53}Ga_{0.47}As$ sidewalls are (111) surfaces.

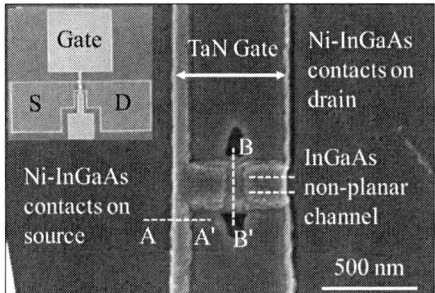

Fig. 2. SEM image shows top view of a multiple-gate $In_{0.53}Ga_{0.47}As$ channel MOSFET having Ni-InGaAs contacts formed on n^{++} $In_{0.53}Ga_{0.47}As$ S/D. The Ni-InGaAs contacts are aligned to the gate electrode. The layout of the device was shown in the inset.

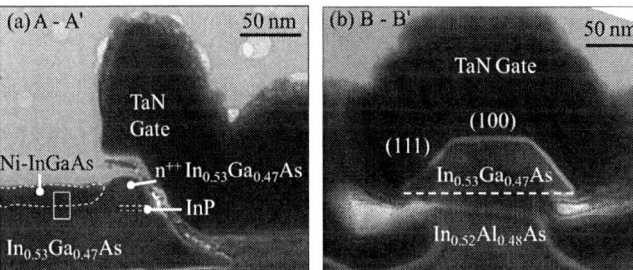

Fig. 3. (a) TEM image shows the device cross-section along A - A' in Fig. 2. Ni-InGaAs contacts, which appear as a darker layer, were formed on n^{++} $In_{0.53}Ga_{0.47}As$ and self-aligned to the gate stack. (b) TEM image shows the device cross-section along B - B' (indicated in Fig. 2). The $In_{0.53}Ga_{0.47}As$ channel is in the shape of a trapezoid with a top width of about 80 nm and bottom width of 170 nm. The height of $In_{0.53}Ga_{0.47}As$ channel is about 52 nm.

ECS Transactions, 45 (4) 209-216 (2012)

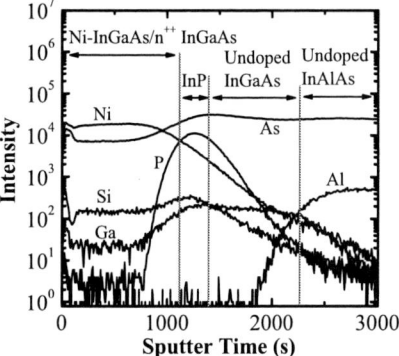

Fig. 4. SIMS profile shows the distribution of elements in device S/D regions and confirms the formation of Ni-InGaAs on n^{++} $In_{0.53}Ga_{0.47}As$ S/D. The vertical grey lines indicated the estimated layer interfaces.

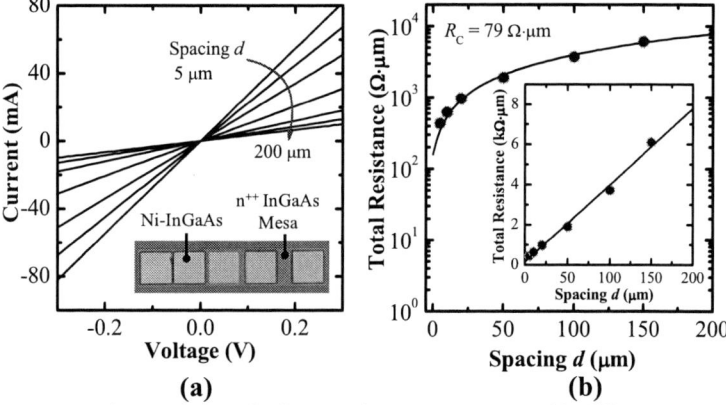

(a)   **(b)**

Fig. 5. (a) *I-V* characteristics of adjacent Ni-InGaAs contacts with different contact spacing d (from 5 to 200 µm) show excellent ohmic behavior. The inset illustrates the layout of the TLM structure. (b) Plot of total resistance between two Ni-InGaAs contacts as a function of the contact spacing (logarithmic scale). The inset shows the total resistance versus spacing d in linear scale. Contact resistance was extracted to be about 79 Ω·µm.

Secondary Ion Mass Spectroscopy (SIMS) characterization was performed in the S/D regions (Fig. 4). The elemental distribution confirms the formation of Ni-InGaAs on n^{++} $In_{0.53}Ga_{0.47}As$ S/D. The vertical grey lines indicated the estimated layer interfaces. *I-V* characteristics of adjacent Ni-InGaAs contacts with different contact spacing d (from 5 to 200 µm) show excellent ohmic behavior [Fig. 5. (a)]. The inset illustrates the layout of the TLM structure. Fig. 5 (b) plots total resistance between two Ni-InGaAs contacts as a

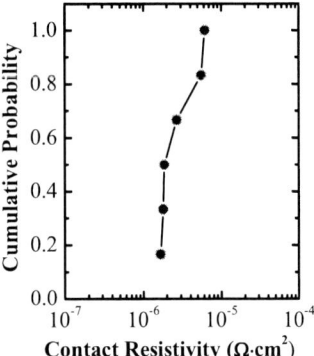

Fig. 6. Cumulative plot showing a tight distribution of contact resistivity measured from TLM structures. The contact resistivity is in the order of 1×10^{-6} $\Omega \cdot cm^2$.

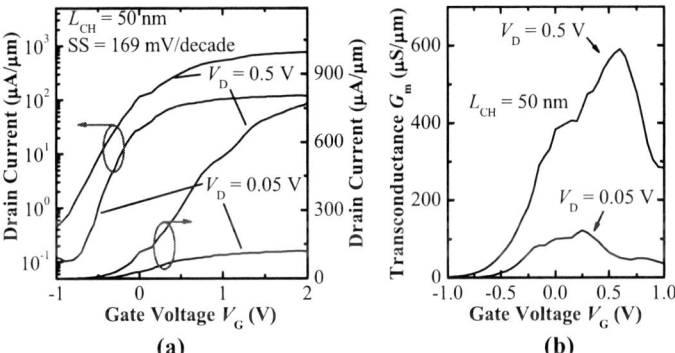

Fig. 7. (a) I_D–V_G curves of a multiple-gate $In_{0.53}Ga_{0.47}As$ MOSFET with channel length L_{CH} of 50 nm, showing good transfer characteristics. The device has a subthreshold swing of 169 mV/decade. (b) G_m–V_G shows a peak extrinsic transcondutance G_m of 590 $\mu S/\mu m$ at $V_D = 0.5$ V.

function of the contact spacing d in logarithmic scale. The inset shows the total resistance versus spacing d in linear scale. Contact resistance was extracted to be about 79 Ω μm. Statistical plot in Fig. 6 shows a tight distribution of contact resistivity measured from TLM structures. Ni-InGaAs shows low contact resistance on n[++] $In_{0.53}Ga_{0.47}As$ and the contact resistivity is in the order of 1×10^{-6} $\Omega \cdot cm^2$.

Fig. 7 (a) shows the I_D–V_G plot of a multiple-gate $In_{0.53}Ga_{0.47}As$ transistor with channel length L_{CH} of 50 nm. The device exhibits good transfer characteristics with substhreshold swing of 169 mV/decade. G_m–V_G in Fig. 7 (b) indicates peak G_m of 590 $\mu S/\mu m$ at $V_D = 0.5$ V. I_D–V_D [Fig. 8 (a)] of the same device in Fig. 7 shows good saturation and pinch-off characteristics. The device shows a drive current exceeding 411 $\mu A/\mu m$ at $V_D = 0.7$ V and $V_G = 0.7$ V. I_G–V_G of the device is plotted in Fig. 8 (b), showing low gate leakage current. The current is normalized by the device width and 3 to 4 orders of magnitude lower than the drive current. This confirms the successful wet etching of the unreacted metal.

Total resistance in linear regime ($V_D = 0.05$ V) at large $V_G = 5$ V indicates low R_{SD} of 364 $\Omega \cdot \mu m$ (Fig. 9). $R_T = R_{SD} + L[W\mu C_{ox}(V_G\text{-}V_T)]^{-1}$ was used to fit the data points, and the fitted curve was extrapolated to a large $V_G = 5$ V to obtain R_{SD}. The plot in Fig. 10

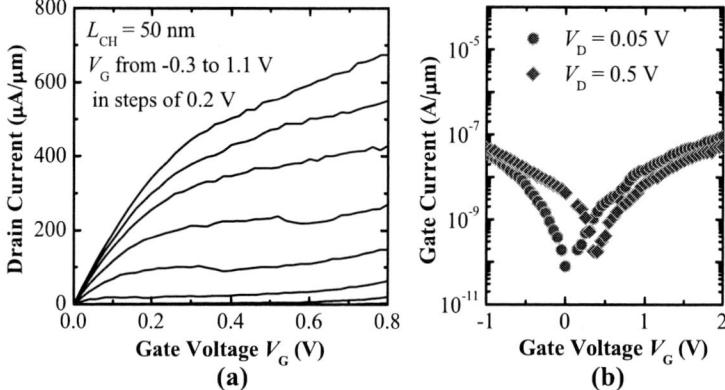

Fig. 8. (a) I_D –V_D of the same device in Fig. 7 shows good saturation and pinch-off characteristics with drive current of over 411 $\mu A/\mu m$ at $V_D = 0.7$ V and $V_G = 0.7$ V. (b) I_G-V_G of the device shows low gate leakage current. The current is normalized by device width and 3 to 4 orders of magnitude lower than the drive current.

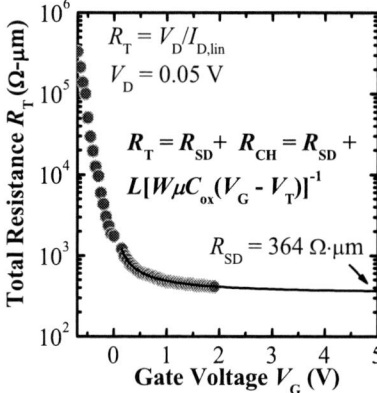

Fig. 9. Total resistance in linear regime ($V_D = 0.05$ V) at large $V_G = 5$ V indicates low R_{SD} of 364 $\Omega \cdot \mu m$. $R_T = R_{SD} + L[W\mu C_{ox}(V_G\text{-}V_T)]^{-1}$ was used to fit the data points, and the fitted curve was extrapolated to a large V_G to obtain R_{SD}.

shows the estimated component elements of the source resistance ($R_{SD} = 2R_S = 2R_D = 364$ $\Omega\cdot\mu m$). Ni-InGaAs resistance (R_{metal}) is calculated to be about 17 $\Omega\cdot\mu m$. n^{++} In$_{0.53}$Ga$_{0.47}$As between Ni-InGaAs contacts and In$_{0.53}$Ga$_{0.47}$As channel gives resistance R_{cap} of about 34 $\Omega\cdot\mu m$. InP barrier resistance ($R_{barrier}$) and Ni-InGaAs contact resistance (R_C) are about 52 and 79 $\Omega\cdot\mu m$, which are the dominant resistance in this self-aligned FinFET structure. Reduction of InP (2 nm) thickness could further reduce the barrier resistance. Another option is to employ new contact materials such as non-alloy Molybdenum which has been reported to give even lower contact resistance (7 $\Omega\cdot\mu m$) on n^{++} In$_{0.53}$Ga$_{0.47}$As [12],[16].

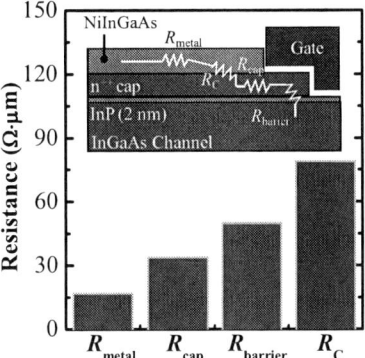

Fig. 10. Estimated components of the source resistance ($R_{SD} = 2R_S = 2R_D = 364$ $\Omega\cdot\mu m$). InP barrier resistance ($R_{barrier}$) and Ni-InGaAs contact resistance (R_C) are the dominant resistance in this self-aligned FinFET structure.

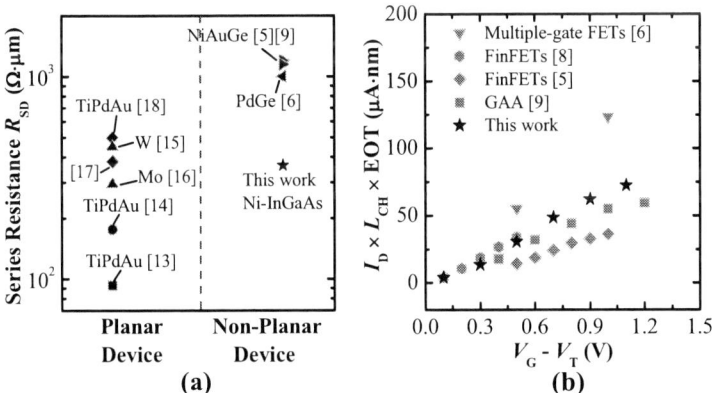

Fig. 11. (a) Much lower R_{SD} (364 $\Omega\cdot\mu m$) was obtained in this work as compared with reported InGaAs non-planar devices with non-self-aligned contacts. (b) $I_D \times L_{CH} \times EOT$ versus overdrives (V_G-V_T) of InGaAs non-planar devices reported in the literature as well as the devices in this work.

Fig. 11 (a) benchmarks R_{SD} of this work with reported InGaAs planar and non-planar devices. Much lower R_{SD} was obtained in this work as compared with other reported InGaAs non-planar devices. This is due to self-alignment of the Ni-InGaAs contact as well as its low contact resistance. Fig. 11 (b) shows $I_D \times L_{CH} \times$ EOT versus overdrives ($V_G - V_T$) of InGaAs non-planar devices reported in the literature as well as the devices in this work.

Conclusion

Multiple-gate $In_{0.53}Ga_{0.47}As$ channel n-MOSFETs with self-aligned Ni-InGaAs contacts were demonstrated for the first time. The device with channel length of 50 nm demonstrates good transfer and output characteristics. Multiple-gate $In_{0.53}Ga_{0.47}As$ MOSFETs with self-aligned Ni-InGaAs contacts formed on *in-situ* doped n^{++} $In_{0.53}Ga_{0.47}As$ S/D exhibit low R_{SD} of 364 $\Omega \cdot \mu m$, due to the self-alignment of Ni-InGaAs contacts as well as the low contact resistance of Ni-InGaAs.

Acknowledgement

Research grant from the National Research Foundation (NRF) (Award number NRF-RF2008-09) is acknowledged.

References

[1] Y. Xuan et al., *IEDM Tech. Dig.* p. 371 (2008).
[2] D.-H. Kim et al., *IEDM Tech. Dig.* p. 719 (2008).
[3] H.-C. Chin et al., *Symp. VLSI Tech. Dig.* p. 244 (2009).
[4] M. Radosavljevic et al., *IEDM Tech. Dig.* p. 319 (2009).
[5] Y. Q. Wu et al., *IEDM Tech. Dig.* p. 331 (2009)
[6] H.-C. Chin et al., *IEEE Elect. Dev. Lett.,* **32**, 146 (2011)
[7] M. Radosavljevic et al., *IEDM Tech. Dig.* p. 126 (2010).
[8] M. Radosavljevic et al., *IEDM Tech. Dig.* p. 765 (2011)
[9] J. J. Gu et al., *IEDM Tech. Dig.* p. 769 (2011)
[10] X. Zhang et al., *Symp. VLSI Tech. Dig.* p. 233 (2010)
[11] X. Zhang et al., *Electrochem. Solid-State Lett.,* **14**, H60 (2011)
[12] U. Singisetti et al., *IEEE Elect. Dev. Lett.,* **30**, 1128 (2009)
[13] Y. Yonai et al., *IEDM Tech. Dig.* p. 307 (2011)
[14] M. Egard et al., *IEDM Tech. Dig.* p. 303 (2011)
[15] N. Waldron et al., *IEDM Tech. Dig.* p. 633 (2007)
[16] T.-W. Kim et al., *IEDM Tech. Dig.* p. 696 (2010)
[17] Y. Sun et al., *IEDM Tech. Dig.* p. 367 (2008)
[18] R. Terao et al., *Appl. Phys. Express,* **4**, 054201 (2011)

ECS Transactions, 45 (4) 217-229 (2012)
10.1149/1.3700471 ©The Electrochemical Society

Sub-100nm Non-planar 3D InGaAs MOSFETs: Fabrication and Characterization

Jiangjiang J. Gu, and Peide D. Ye

Department of Electrical and Computer Engineering and Birck Nanotechnology Center,
Purdue University, West Lafayette, Indiana 47907, USA

InGaAs MOSFETs have been considered promising candidate for
post-Si logic devices beyond 14nm technology node. To meet the
increasing demand in electrostatic control at sub-100nm channel
lengths, non-planar 3D structures have been introduced to the
fabrication of InGaAs MOSFETs. In this paper, the fabrication and
characterization of various non-planar 3D InGaAs MOSFETs have
been demonstrated and summarized, including InGaAs
heterostructure FinFETs, InGaAs-on-nothing nanowire MOSFETs,
and InGaAs gate-all-around nanowire MOSFETs. It is shown that
the implementation of 3D structure greatly reduces short channel
effect and improves scalability of InGaAs MOSFETs. The gate-all-
around nanowire structure has been fabricated by a novel top-down
approach for the first time and is found to offer great scalability
down to at least 50nm channel length with good transport property,
making InGaAs gate-all-around nanowire MOSFETs strong
candidate for ultimately scaled III-V logic technology.

Introduction

InGaAs MOSFETs with encouraging device performance have been demonstrated,
and are considered promising candidate to replace Si CMOS at 14nm technology node
and beyond [1-3]. Much attention has been focused on the interface engineering between
high-k dielectric and III-V channel to reduce the interface trap density (D_{it}), which is
critical for realizing steep sub-threshold swing (SS) at off-state as well as large current
drive at on-state. However, the effective control of short channel effect by utilizing
advanced device structure such as non-planar 3D or ultra-thin body becomes equally
important as device dimension reaches sub-100nm regime. As Si CMOS manufacturing
enters a new era with the "tri-gate" design at 22 nm node [4], similar non-planar 3D
approach is yet to be explored on its III-V counterpart [5]. In this paper, we summarize
and review our recent development on several InGaAs non-planar 3D architecture,
including, InGaAs heterostructure FinFET (HFinFET) [6], InGaAs-on-nothing nanowire
MOSFETs (InGaAsON FET) [7], and InGaAs gate-all-around nanowire MOSFETs
(InGaAs GAA FET) [8].

InGaAs HFinFETs

InGaAs surface-channel FinFETs have been demonstrated to suppress short channel
effects and extend the scaling of InGaAs MOSFETs down to 100nm [5][9]. On the other
hand, the buried-channel InGaAs MOSFETs with GaAs/AlGaAs, InAlAs or InP barrier
[10-13] and quantum-well FETs (QWFETs) with thin InP barrier [2] have been shown to
offer higher transconductance (g_m), higher effective mobility and lower SS compared to
surface-channel III-V MOSFETs. Therefore, an InGaAs FinFET with wide bandgap

217

barrier layers would benefit from both the good short channel effect control of the FinFET and the low voltage operation of the heterostructure design [14]. In this work, we demonstrated InGaAs HFinFETs with different Fin width (W_{Fin}) and two channel thickness of 10nm (type I) and 60nm (type II) shown in Figure 1 and 2.

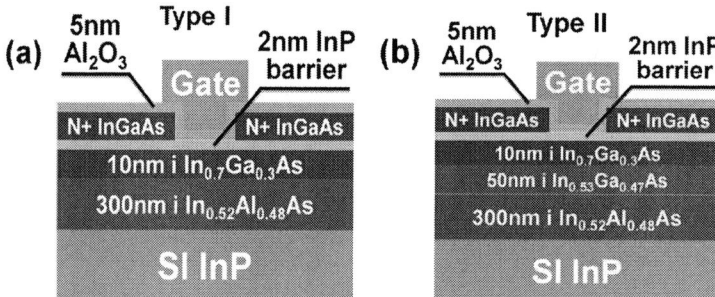

Figure 1. Schematic diagram of InGaAs HFinFETs with (a) 10nm channel (b) 60nm channel.

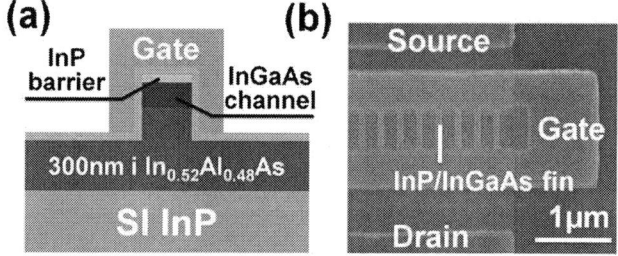

Figure 2. (a) Schematic cross section of InGaAs HFinFETs. (b) Top-view SEM image of a InGaAs HFinFET (type I) with W_{Fin}=30nm.

MOSFET fabrication started with a 2 inch semi-insulating InP substrate. A 300 nm undoped $In_{0.52}Al_{0.48}As$ buffer layer, 50 nm undoped $In_{0.53}Ga_{0.47}As$ channel layer (type I only), 10 nm undoped $In_{0.7}Ga_{0.3}As$ channel layer (type I and type II), 2 nm undoped InP barrier layer and 20 nm N+ doped InGaAs layer were sequentially grown by molecular beam epitaxy. Device isolation and gate recess etching were then performed using citric acid based solution. The gate lengths of the devices were varied from 0.5μm down to 100 nm. For non-planar devices, a fin etching process was done using BCl_3/Ar based reactive ion etching [5]. The smallest W_{Fin} defined was 30 nm. After short buffered oxide etch (BOE) dip, the samples were soaked in 10% $(NH_4)_2S$. The passivation time was fixed at 10 minutes. Previous Al_2O_3/InP capacitance-voltage studies revealed that 10 minute treatment was sufficient to achieve an effective passivation. The air exposure after sulfur treatment was minimized. The samples were then loaded into an ASM F-120 atomic-layer deposition (ALD) reactor for 5 nm Al_2O_3 deposition at 300 °C. Source/drain contacts were then formed by Au/Ge/Ni deposition and 350 °C rapid thermal annealing process (RTA). Finally, Ni/Au was electron beam evaporated as gate metal. Since sulfur passivation was found to be unstable after thermal treatment higher than 400 °C [15], no post deposition annealing (PDA) was performed after ALD gate dielectric deposition, and the thermal budget of the entire fabrication process was as low as 350 °C. All patterns were defined by a Vistec UHR electron beam lithography system.

Electrical characterization of InGaAs HFinFET with 10nm channel (Type I)

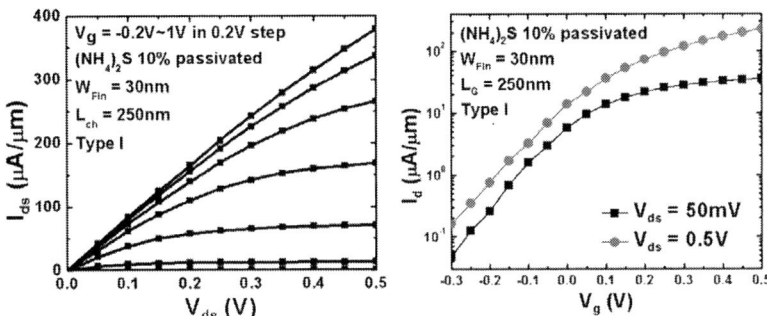

Figure 3. I_{ds}-V_{ds} and I_{ds}-V_{gs} of InGaAs HFinFET (Type I) with W_{Fin} =30nm and L_{ch}=250nm.

Figure 3 show the well-behaved output and transfer characteristics of a InGaAs HFinFET with L_{ch}= 250 nm, W_{Fin} = 30 nm and passivated with 10% $(NH_4)_2S$ before gate oxide deposition. A saturation drain current of 380 μA/μm and g_m of 557 μS/μm is obtained at a V_{ds} = 0.5 V. The threshold voltage (V_T) of the device is -0.05V from linear extrapolation at V_{ds} = 50 mV and -0.18V using 1μA/μm metric at V_{ds} = 0.5 V. A SS of 120 mV/dec and DIBL of 99 mV/V are also achieved. Compared to deep-submicron surface-channel InGaAs MOSFETs, a higher g_m is obtained at a lower drain voltage with the same gate oxide thickness [16], indicating the advantage of buried-channel devices for low-voltage operation.

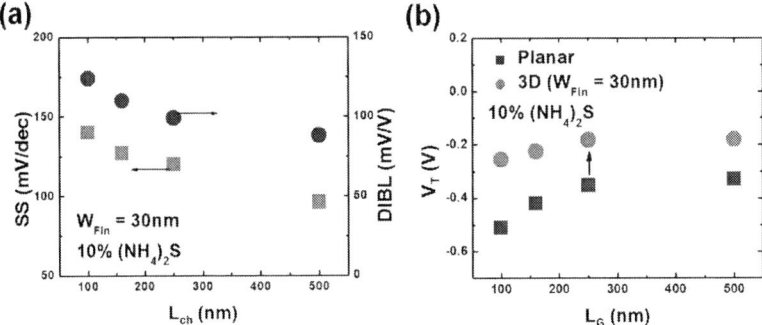

Figure 4. (a) SS and DIBL versus L_{ch} of 3D (W_{Fin} = 30 nm) InGaAs HFinFETs (Type I) (b) V_T versus L_G of planar and 3D (W_{Fin} = 30 nm) InGaAs HFinFET (Type I). V_T is determined by 1 μA/μm metric at V_{ds} = 0.5 V.

Furthermore, we investigate the scaling metrics of the InGaAs HFinFETs with 10% $(NH_4)_2S$ passivation and the gate lengths of the devices varied from 0.5 μm down to 100 nm. Figure 4 (a) shows the SS and DIBL versus L_{ch} for non-planar devices with W_{Fin} = 30 nm, where SS is obtained at a drain voltage of 0.5 V. It is found that SS and DIBL gradually increase with L_G shrinking due to the SCE. Further suppression of SCE can be achieved by reducing W_{Fin} [5], decreasing the equivalent oxide thickness, or implementing more advanced 3D structure such as gate-all-around structure [8]. Figure 4 (b) shows the V_T versus L_{ch} for planar and 3D devices with $W_{Fin} = 30$ nm. A 0.15V to 0.25V positive V_T shift has been observed for 3D devices, making the device operation more approaching enhancement-mode. Moreover, 3D devices show better threshold roll-off property due to a better electrostatic control of the channel. These results highlight the importance of introducing advanced 3D structure to the fabrication of III-V MOSFETs at deep submicron gate lengths.

Electrical characterization of InGaAs HFinFET with 60nm channel (Type II)

Figure 5 (a) show the transfer characteristics of planar and HFinFET with 60nm channel. Due to the thicker channel, type II planar devices show much worse off-state metrics than that of type I devices. Therefore, the non-planar 3D structure plays a more important role in this type of devices. Figure 5 (b) show the SS scaling metrics as a function of L_{ch}. The planar devices suffer from severe short channel effects, while the HFinFET is highly resistant to short channel effects. The SS of HFinFET with 60nm channel thickness is a constant regardless of the L_{ch}, showing good electrostatic control with L_{ch} down to at least 10nm. It is noted that the type II HFinFET show better gate control than type I due to the higher H_{Fin}/W_{Fin} ratio of the devices. To benefit fully from the 3D FinFET structure, fins of high aspect ratio is strongly needed.

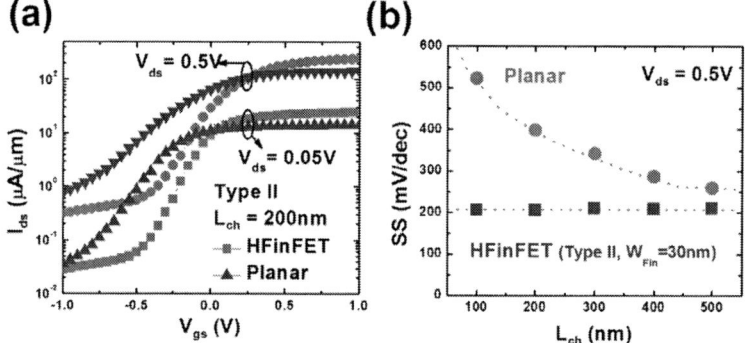

Figure 5. (a) Transfer characteristics of planar and HFinFET (Type II) with 60nm channel (b) SS as a function of L_{ch} for planar and HFinFET (Type II) at V_{ds}=0.5V.

InGaAs-On-Nothing nanowire MOSFETs

Silicon on nothing (SON) MOSFETs have been proposed [17] and experimentally demonstrated [18], providing a solution for quasi-total suppression of short channel effects and DIBL with L_{ch} down to 30 nm. The key fabrication processes for SON MOSFETs involve selective removal of the SiGe layer and formation of an air tunnel underneath the Si channel. Similarly, a SON counterpart on the III-V platform, namely III-V on nothing (III-VON), can also be realized if a selective etching process is developed to form an air gap underneath the III-V channel. We report the first experimental demonstration of III-VON MOSFETs with $In_{0.53}Ga_{0.47}As$ as the channel and atomic-layer-deposited (ALD) Al_2O_3 as the gate dielectric. The starting substrate is 30nm p- $In_{0.53}Ga_{0.47}As$ epitaxially grown on heavily p-doped InP substrate. The key fabrication has been enabled by well-controlled HCl based selective etching of InP over InGaAs. It is found that the InGaAs fins have to be patterned along [010] direction in order to achieve a successful channel release process, owing to the anisotropy of InP wet etching. III-VON MOSFETs with gate length down to 50 nm and fin width down to 40 nm were fabricated. A low DIBL of 45mV/V at 50 nm gate length is observed experimentally, which is consistent with the numerical simulation results from 3D Synopsys Sentaurus TCAD. This shows that the III-VON structure is effective at suppressing the SCE of III-V MOSFETs down to at least L_G = 50 nm.

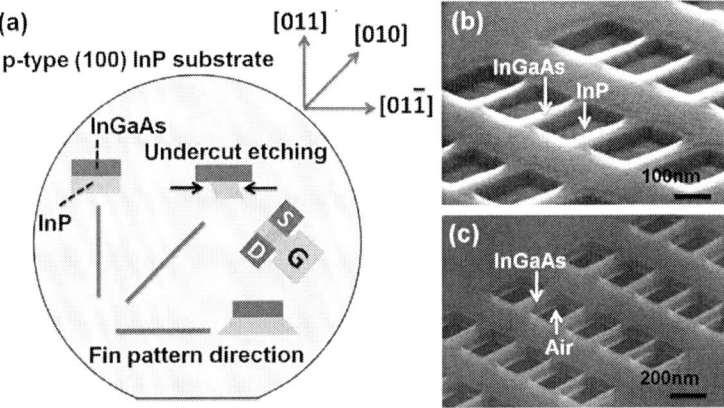

Figure 6 (a) Schematic diagram of fin patterning direction, release etching profile, device alignment to the substrate (b) Tilted SEM image of fin test structures after fin dry etching and before wire release (c) Tilted SEM image of fin testing structures after wire release.

MOSFET fabrication started with a 2 inch p+ InP wafer. A 30nm $In_{0.53}Ga_{0.47}As$ layer with p-doping of 2×10^{16} cm^{-3} was grown on InP substrate by molecular beam epitaxy as the channel layer. After surface treatment with $(NH_4)OH$ solution, a 10nm Al_2O_3 was grown by ALD as an encapsulation layer. Source/drain regions were then defined and Si ion implantation at energy of 20 keV and dose of 1×10^{14} cm^{-2} was performed. The shortest gate length of 50 nm was defined by the separation between source and drain regions. Dopant activation was done using rapid thermal annealing (RTA) at 600°C for 15 seconds in nitrogen ambient. Next, the InGaAs fins were patterned along [010] direction as shown in Figure 1 (a), using diluted ZEP520A electron-beam resist with a thickness of 200 nm. The fin etching was performed with BCl_3/Ar gas using a Panasonic high density plasma etcher. After removing the resist, the sample was treated sequentially with diluted buffered oxide etch (BOE) and diluted mixture of HCl and hydrogen peroxide (H_2O_2). Then the channel release process was performed using $HCl:H_2O$ (1:2) solution. It is known that HCl based solution has high selectivity between InP and InGaAs. However, the InP etching is found to be highly anisotropic, and undercut etching is only possible along <100> directions [19]. Test fin structures along [011], [010] and [01-1] were patterned, followed by etching in diluted HCl. For different fin patterning directions, the etching profile varies as depicted in Figure 6 (a). Figure 6 (b) and (c) show the tilted scanning electron microscopy (SEM) images for fin test structures before and after channel release process respectively, where successful undercut etching was demonstrated. Patterning the fin along <100> directions is the key to realize the III-V channel release at the deep sub-micron scale. After channel release, a 5 nm Al_2O_3 was regrown using ALD as the gate dielectric. Note that for the purpose of process demonstration, no pre-gate surface passivation such as $(NH_4)_2S$ treatment was carried out here. The gate structure was formed by electron beam evaporation of Ni/Au and liftoff process. Due to the vertical directionality of the evaporation, the air gap naturally remains underneath the channel. Au/Ge/Ni source/drain metal was then evaporated and annealed at 350°C in nitrogen ambient. Finally, test pads were deposited which concluded the fabrication processes. All patterns were defined by a Vistec VB-6 UHR electron beam lithography system.

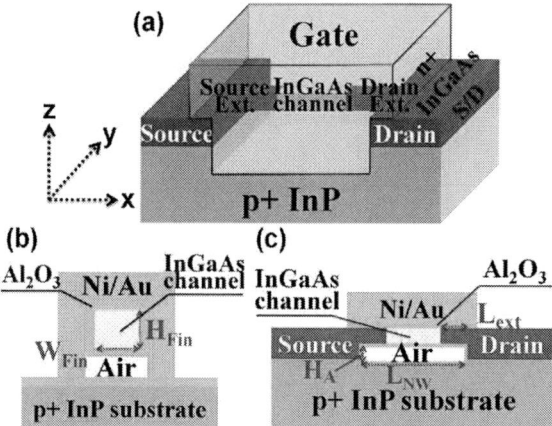

Figure 7 (a) Schematic diagram of a III-VON MOSFET with $In_{0.53}Ga_{0.47}As$ channel and Al_2O_3 gate dielectric from a bird's eye view (b) Cross sectional view of a III-VON MOSFET in x-y plane (c) Cross sectional view of a III-VON MOSFET in x-z plane.

Figure 7 (a) shows the schematic diagram of a finished III-VON device from a bird's eye view. Figure 7 (b) and (c) depict the schematic cross section of the device in the y-z plane and the x-z plane, respectively. The fin height ($H_{Fin} = 30$ nm) is determined by the initial InGaAs layer thickness. The smallest W_{Fin} achieved is 40 nm. The nanowire length (L_{NW}) in this work is fixed at 300 nm, yielding a source/drain extension length (L_{ext}) of around 125 nm for a 50 nm L_{ch} device. The smallest air gap height (H_A) is around 40 nm, controlled by the release etching time. The L_{ch} of the devices vary from 100 nm down to 50 nm. Figure 8 (a) and (b) shows the output, transfer characteristic and gate leakage current versus gate voltage of a typical III-VON MOSFET with L_{ch} of 50 nm, W_{Fin} of 40 nm, and four wires in parallel. To better evaluate the intrinsic device performance, source current is used to eliminate the effect from non-ideal source/drain junction leakage current. The device operates in enhancement mode, with a threshold voltage of 0.36 V from linear extrapolation at a drain voltage of 50 mV. A low DIBL of 45 mV/V is obtained at the shortest gate length of 50 nm, thanks to the III-VON structure. As a comparison, a 100 nm gate length InGaAs FinFET has a DIBL of around 180 mV/V [5]. The subthreshold swing (SS) is found to be around 200 mV/dec, indicating relatively large interface trap density (D_{it}). Surface treatment before the formation of ALD Al_2O_3 gate dielectric which could have improved the D_{it} was not performed since that was not our main purpose in this study. Gate leakage current is similar to that in Ref. [5] and stays very low at gate voltage less than 1V. The saturation current at a drain voltage of 1.6V and a gate voltage of 2V reaches 10μA/μm, normalized by the total width of the gated region, i.e. $W_{ch} = 2 \times H_{Fin} + W_{Fin}$. The current can be further improved by applying $(NH_4)_2S$ pre-gate treatment and reducing source/drain series resistance.

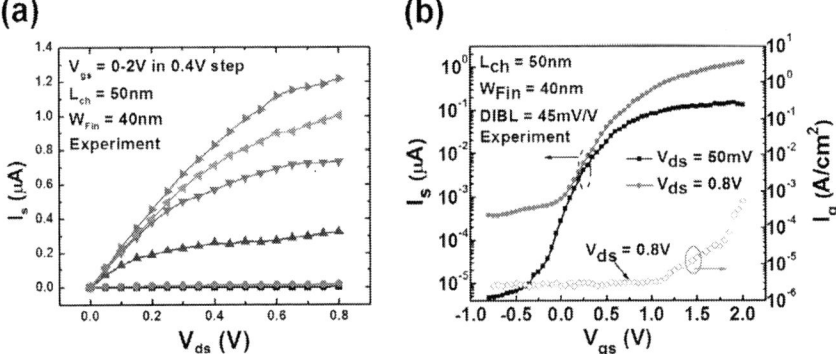

Figure 8 (a) Output and (b) transfer characteristics of InGaAs-ON MOSFETs with L_{ch}=50nm, W_{Fin}=40nm.

To further confirm the experimental data and examine the effects of various design parameters on device performance, 3-dimensonal TCAD simulation was performed using Synopsys Sentaurus. Device structures were first created according to the experimental parameters, i.e. $W_{Fin} = 40$ nm, $H_{Fin} = 30$ nm, etc. The Poisson's equation, electron and hole continuity equations were solved using a coupled solver at each mesh node to obtain various output parameters such as potential, electric field, electron and hole density, etc. No interface traps were incorporated in the simulation. From the simulated transfer characteristics, DIBL was extracted and compared with the experimental data as shown in Figure 9. The experimental data agrees well with the simulation results. Moreover, the DIBL obtained from InGaAs FinFETs in Ref. [5] is also plotted in the same figure. The III-VON structure shows significant improvement in DIBL reduction, confirming that the more advanced 3D structures are beneficial for the suppression of the SCE.

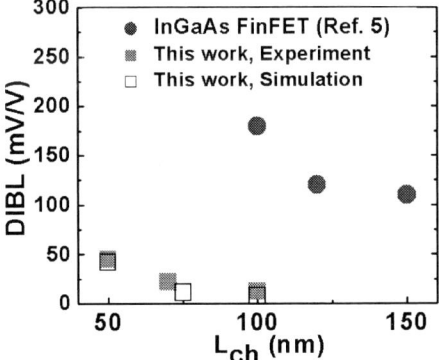

Figure 9 DIBL versus L_{ch} for III-VON MOSFETs from experiment (square) and simulation (hollow square) compared to that of InGaAs FinFET (circle).

InGaAs Gate-all-around nanowire MOSFETs

The GAA structure has been proven on Si CMOS to be the most resistant to SCE, thanks to having the best gate electrostatic control [20-22]. Therefore, a III-V GAA FET is the most promising candidate for the ultimately scaled III-V FETs. We report the first experimental demonstration of inversion-mode $In_{0.53}Ga_{0.47}As$ GAA FETs by a top-down approach with ALD Al_2O_3/WN gate stacks. Benefiting from the GAA structure, we have demonstrated the shortest $L_{ch} = 50nm$ III-V MOSFETs to date with well-behaved on- and off-state characteristics. A systematic scaling metrics study has been carried out for $In_{0.53}Ga_{0.47}As$ GAA FETs with L_{ch} from 110nm down to 50nm, W_{Fin} of 30nm and 50nm, fin height (H_{Fin}) of 30nm, and wire lengths L_{NW} of 150 to 200nm. The SCE has been effectively suppressed by the advanced 3D design.

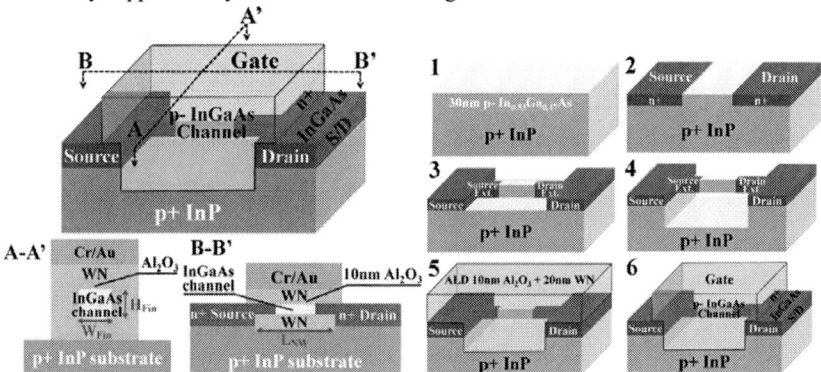

Figure 10 Schematic diagram of InGaAs GAA FETs from bird's eye view.

Figure 11 Key Fabricatin process of InGaAs GAA FETs.

Fig. 10 shows a schematic view of an $In_{0.53}Ga_{0.47}As$ GAA FET fabricated in this work. Fig. 11 depict the key fabrication processes for $In_{0.53}Ga_{0.47}As$ GAA FETs. A 30nm p-doped 2×10^{16} cm^{-3} $In_{0.53}Ga_{0.47}As$ channel layer was epitaxially grown on a p+ (100) InP substrate by MBE as the starting material (Fig. 11-1). After surface degrease and NH_4OH pretreatment, 10nm Al_2O_3 was grown by ALD as an encapsulation layer. Source/drain Si implantation was then performed at an energy of 20keV and a dose of 1×10^{14} cm^{-2} (Fig. 11-2). The dopant activation was carried out at 600 °C for 15 seconds in nitrogen ambient. The source/drain separation determines the final L_{ch} of the devices. After removing the encapsulation layer by buffered oxide etch (BOE), the InGaAs fin etching was done by BCl_3/Ar high density plasma etching (HDPE) (Fig. 11-3). The diluted ZEP520A electron-beam resist with a thickness of 100 nm was used as a hard mask for the fin etching and the smallest W_{Fin} defined was 30nm. After surface cleaning in BOE and diluted $HCl:H_2O_2$ solution, the InGaAs channel release process was carried out using $HCl:H_2O$ (1:2) solution (Fig. 11-4). HCl based solution can selectively etch InP over InGaAs. However, the etching is found to be highly anisotropic. Therefore the InGaAs fins have to be patterned along <100> directions for a successful release process. Fig. 12 (a) shows the cross-sectional STEM image of InGaAs nanowire test structures wrapped by 50nm ALD Al_2O_3 on InP substrate, confirming the nanowires are completely released. After channel release, the samples were soaked in 20% $(NH_4)_2S$ for pre-gate interface passivation. Then the samples were immediately transferred to an ASM F-120 ALD reactor via room ambient. 10nm Al_2O_3 was regrown as the gate dielectric at 300 °C.

20nm WN metal gate was then deposited in a separate ALD reactor at 385 °C (Fig. 11-5), with a resistivity of ~4000μΩ·cm [23]. The conformal deposition of ALD Al_2O_3/WN surrounding the nanowire channel is the key fabrication process for realizing the GAA structure. After gate stack deposition, gate etch process was performed using CF_4/Ar HDPE, where Cr/Au gate pattern was defined as the hard mask (Fig. 11-6). The CF_4 based dry etching chemistry provides excellent selectivity between WN and Al_2O_3, resulting in a damage-free gate oxide. The source/drain contact was then formed by electron beam evaporation of Au/Ge/Ni, followed by 350 °C rapid thermal annealing in nitrogen ambient. Finally, the Ti/Au test pads were defined. The fabricated MOSFETs have a nominal L_{ch} varying from 50nm to 120nm, W_{Fin} from 30nm to 50nm, and different numbers of parallel channels (1 wire, 4 wires, 9 wires or 19 wires). Fig. 12 (b) shows the SEM image of a InGaAs GAA FET with 4 parallel wires.

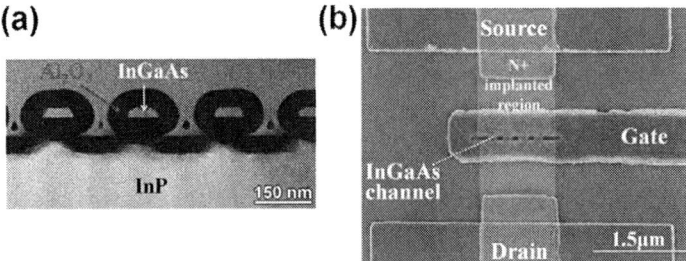

Figure 12 (a) Cross-sectional STEM image of InGaAs nanowire test structures wrapped by 50nm ALD Al_2O_3 on InP substrate (b) Top view SEM image of a finished InGaAs GAA FET with 4 parallel wires of W_{Fin} = 30nm , L_{NW} = 200nm and L_{ch} = 50nm.

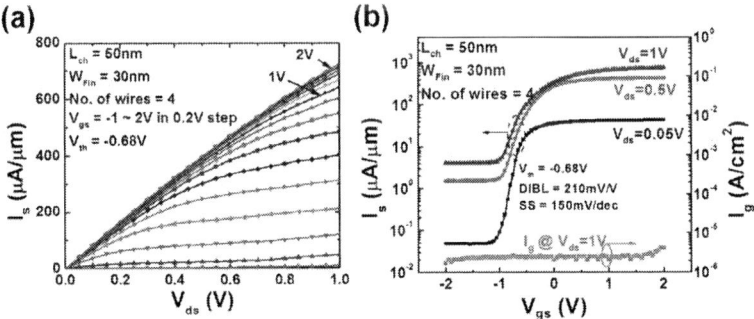

Figure 13 (a) output and (b) transfer characteristics of InGaAs GAA FET with L_{ch}=50nm, W_{NW}=30nm.

Fig. 13 (a) and (b) show the well-behaved output and transfer characteristics as well as I_g-V_g of a L_{ch} = 50nm GAA FET. The current here is normalized by the total perimeter of the $In_{0.53}Ga_{0.47}As$ channel, i.e. $W_G = (2W_{Fin} + 2H_{Fin}) \times$ (No. of wires). A representative 50nm L_{ch} device shows on-current of 720μA/μm, transconductance of 510μS/μm and reasonable off-state characteristics with subthreshold swing (SS) of 150mV/dec and drain-induced barrier lowering (DIBL) of 210mV/V. Although operating in inversion-mode, the threshold voltage of the device is -0.68V from linear extrapolation at V_{ds}=50mV due to the relatively low work function of ALD WN metal (~4.6eV). Due to the junction leakage current and a very large area ratio (>10³) between implanted junction and GAA channels, the source current is used to obtain the intrinsic current in the channel. Gate leakage current is minimal in the entire gate voltage range, indicating 10nm Al₂O₃ is sufficient for GAA structure and further equivalent oxide thickness (EOT) scaling is achievable. It also shows that the WN gate etch process is damage-free.

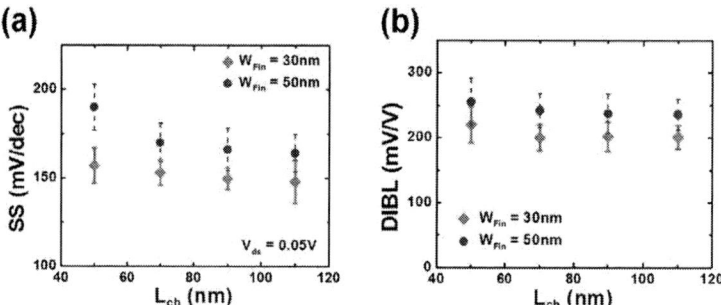

Figure 14 (a) SS and (b) DIBL scaling metrics of InGaAs GAA FETs with L_{ch} down to 50nm.

Fig. 14 show the off-state (SS and DIBL) scaling metrics for L_{ch} = 50 - 110nm with W_{Fin}=30nm and 50nm. The SS for 30nm W_{Fin} devices are almost unchanged at around 150mV/dec when scaling L_{ch} down to 50nm, indicating excellent control of SCE, whereas the 50nm W_{Fin} devices show larger SS, which increases with scaling of L_{ch}. It is noted here that the 100nm L_{ch} InGaAs FinFET with 5nm Al₂O₃ gate oxide shows similar SS [5] as the 50nm L_{ch} GAA FET with 10nm Al₂O₃ in this work. This translates to at least a factor of 2 improvement of midgap D_{it} (~5.6×10¹²/cm²·eV) achieved. The improved interface quality indicates that the newly-developed channel release process can provide a smooth damage-free InGaAs bottom surface. Fig. 11 shows that 30nm W_{Fin} devices have smaller DIBL and the DIBL is roughly independent of L_{ch}, confirming the effective SCE control. Further SS and DIBL reduction can be achieved by scaling down EOT and reducing the InGaAs nanowire dimension. Figure 15 compares g_m EOT product for InGaAs GAA FETs with InGaAs MOSFETs fabricated in our group recently. The InGaAs GAA FETs benefits from the continuous scaling of the L_{ch} down to 50nm.

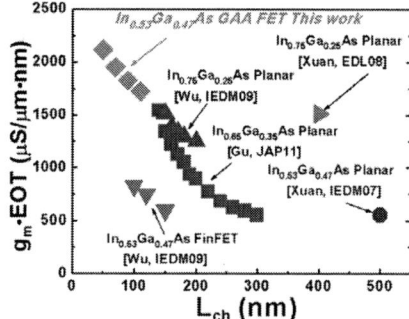

Figure 15 Benchmarking g_m·EOT of planar and non-planar InGaAs surface-channel MOSFETs.

We have demonstrated for the first time inversion-mode $In_{0.53}Ga_{0.47}As$ GAA MOSFETs with ALD Al_2O_3/WN gate stacks. The highest saturation current reaches 1.17mA/µm at L_{ch} = 50nm and V_{ds} = 1V with $g_{m,max}$ = 701µS/µm. Detailed scaling metrics study shows that the 3D GAA structure can effectively control the SCE with L_{ch} scaling down to at least 50nm, making III-V GAA FET a very promising candidate for ultimately scaled III-V logic device technology.

Summary

In this paper, we have demonstrated non-planar 3D InGaAs MOSFETs including InGaAs HFinFETs, InGaAs-on-nothing nanowire MOSFETs and InGaAs gate-all-around nanowire MOSFETs. InGaAs HFinFETs with high aspect ratio fins show better scalability and extend the scaling of InGaAs MOSFETs down to 100nm L_{ch}. The introduction of nanowire structure further increases the resistance to short channel effects and record-low DIBL of 45mV/V has been achieved at 50nm L_{ch} for InGaAs-ON MOSFETs. The InGaAs GAA FETs show the best electrostatic control and is the most promising candidate for future high-speed low-power logic applications. Further interface passivation, EOT scaling and source/drain engineering is needed to optimized the device performance of the InGaAs GAA FETs.

Acknowledgments

The authors would like to thank R. Wang, M. Luisier, D. A. Antoniadis, M. S. Lundstrom, R. G. Gordon, X. L. Li, and J. del Alamo for the valuable discussions and Y. Q. Liu, C. Zhang, L. Dong for technical assistance. The work is supported in part by NSF and the SRC FCRP MSD Focus Center.

References

1. Y. Xuan, Y. Q. Wu and P. D. Ye, "High-performance inversion-type enhancement-mode InGaAs MOSFET with maximum drain current exceeding 1A/mm," IEEE Electron Device Letters, vol. 29, pp. 294-296, Apr. 2008.
2. M. Radosavljevic, B. Chu-Kung, S. Corcoran, G. Dewey, M. Hudait, J. Faste-nau, J. Kavalieros, W. Liu, D. Lubyshev, M. Metz, K. Millard, N. Mukherjee, W. Rachmady, U. Shah, and R. Chau, "Advanced high-k gate dielectric for high-performance short-channel $In_{0.7}Ga_{0.3}As$ quantum well field effect transistors on silicon substrate for low power logic applications," 2009 IEEE International Electron Devices Meeting, pp. 319-322, Dec. 2009.
3. J. A. del Alamo, "Nanometre-scale electronics with III-V compound semiconductors," Nature, vol. 479, no. 7373, pp. 317–323, Nov. 2011.
4. B. S. Doyle, S. Datta, M. Doczy, S. Hareland, B. Jin, J. Kavalieros, T. Linton, A. Murthy, R. Rios, and R. Chau, "High performance fully-depleted tri-gate CMOS transistors", IEEE Electron Device Letters, vol. 24, pp. 263-265, Apr. 2003.
5. Y. Q. Wu, R. S. Wang, T. Shen, J. J. Gu, and P. D. Ye, "First experimental demonstration of 100 nm inversion-mode InGaAs FinFET through damage-free sidewall etching," 2009 IEEE International Electron Devices Meeting, pp.331-334, Dec. 2009.
6. J. J. Gu, A. T. Neal, and P. D. Ye, "Effects of $(NH_4)_2S$ passivation on the off-state performance of 3-dimensional InGaAs metal-oxide-semiconductor field-effect transistors", Applied Physics Letters, vol. 99, pp. 152113-152115, Oct. 2011.

7. J. J. Gu, O. Koybasi, Y. Q. Wu, and P. D. Ye, "III-V-on-nothing metal-oxide-semiconductor field-effect transistors enabled by top-down nanowire release process: Experiment and simulation", Applied Physics Letters, vol. 99, pp. 112113-112115, Sep. 2011.

8. J. J. Gu, Y. Q. Liu, Y. Q. Wu, R. Colby, R. G. Gordon, and P. D. Ye, "First Experimental Demonstration of Gate-all-around III-V MOSFETs by Top-down Approach", 2011 IEEE International Electron Devices Meeting, pp.769-772, Dec. 2011.

9. H.-C. Chin, X. Gong, L. Wang, H. K. Lee, L. Shi and Y.-C. Yeo, "III-V Multiple-Gate Field-Effect Transistors With High-Mobility $In_{0.7}Ga_{0.3}As$ Channel and Epi-Controlled Retrograde-Doped Fin", IEEE Electron Device Letters, vol. 32, pp. 146-148, Feb. 2011.

10. M. Passlack, P. Zurcher, K. Rajagopalan, R. Droopad, J. Abrokwah, M. Tutt, Y.-B. Park, E. Johnson, O. Hartin, A. Zlotnicka, P. Fejes, R. J. W. Hill, D. A. J. Moran, X. Li, H. Zhou, D. Macintyre, S. Thoms, A. Asenov, K. Kalna, and I. G. Thayne, "High Mobility III-V MOSFETs For RF and Digital Applications", International Electron Device Meeting, pp. 621-614, Dec. 2007.

11. H.C. Lin, T. Yang, H. Sharifi, S.K. Kim, Y. Xuan, T. Shen, S. Mohammadi, and P.D. Ye, "Direct-current and radio-frequency characterizations of GaAs metal-insulator-semiconductor field-effect transistors enabled by self-assembled nanodielectrics", Applied Physics Letters, vol. 91, pp. 092103, 2007.

12. Y. Sun, E. W. Kiewra, J. P. de Souza, J. J. Bucchignano, K. E. Fogel, D. K. Sadana, and G. G. Shahidi, "Scaling of $In_{0.7}Ga_{0.3}As$ buried-channel MOSFETs", International Electron Device Meeting, pp. 367-370, Dec. 2008.

13. H. Zhao, Y.-T. Chen, J. H. Yum, Y. Wang, N. Goel, and J. C. Lee, "High performance $In_{0.7}Ga_{0.3}As$ metal-oxide-semiconductor transistors with mobility >$4400cm^2$/Vs using InP barrier layer", Applied Physics Letters, vol. 94, pp. 193502, May 2009.

14. M. Radosavljevic, G. Dewey, J. M. Fastenau, J. Kavalieros, R. Kotlyar, B. C.-K, W. K. Liu, D. Lubyshev, M. Metz, K. Millard, N. Mukherjee, L. Pan, R. Pillarisetty, W. Rachmady, U. Shah and R. Chau, "Non-planar, multi-gate InGaAs quantum well field effect transistors with high-k gate dielectric and ultra-scaled gate-to-drain/gate-to-source separation for low power logic applications", International Electron Device Meeting, pp. 611-614, Dec. 2010.

15. J. J. Gu, Y. Q. Wu and P. D. Ye, "Effects of gate-last and gate-first process on deep submicron inversion-mode InGaAs n-channel metal-oxide-semiconductor field effect transistors", Journal of Applied Physics, vol. 109, pp. 05709, Mar. 2011.

16. Y. Q. Wu, W. K. Wang, O. Koybasi, D. N. Zakharov, E. A. Stach, S. Nakahara, J. C. M. Hwang and P. D. Ye, "0.8-V Supply Voltage Deep-Submicrometer Inversion-Mode $In_{0.75}Ga_{0.25}As$ MOSFET", IEEE Electron Device Letters, vol. 30, pp. 700-702, Jul. 2009.

17. M. Jurczak, T. Skotnicki, M. Paoli, B. Tormen, J.-L. Regolin, C. Morin, A. Schiltz, J. Martins, R. Pantel and J. Galvier, "SON (silicon on nothing)-a new device architecture for the ULSI era", 1999 Symposium on VLSI Technology, pp.29-30, 1999.

18. S. Monfray, T. Skotnicki, Y. Morand, S. Descombes, M. Paoli, P. Ribot, A. Talbot, D. Dutartre, F. Leverd, Y. Lefriec, R. Pantel, M. Haond, D. Renaud, M.-E. Nier, C. Vizioz, D. Louis and N. Buffet, "First 80 nm SON (Silicon-On-Nothing)

MOSFETs with perfect morphology and high electrical performance", International Electron Device Meeting, pp. 645-648, Dec. 2001.

19. K. Kurishima, S. Yamahata, H. Nakajima, H. Ito and N. Watanabe, "Initial degradation of base-emitter junction in carbon-doped InP/InGaAs HBT's under bias and temperature stress", IEEE Electron Device Letters, vol. 19, pp. 303, 1998.

20. S. D. Suk, S.-Y. Lee, S.-M. Kim, E.-J. Yoon, M.-S. Kim, M. Li, C. W. Oh, K. H. Yeo, S. H. Kim, D.-S. Shin, K.-H. Lee, H. S. Park, J. N. Han, C. J. Park, J.-B. Park, D.-W. Kim, D. Park, and B.-I. Ryu, "High Performance 5nm radius Twin Silicon Nanowire MOSFET(TSNWFET): Fabrication on Bulk Si Wafer, Characteristics, and Reliability", International Electron Device Meeting, pp.717, Dec. 2005.

21. N. Singh, K. D. Buddharaju, S. K. Manhas, A. Agarwal, S. C. Rustagi, G. Q. Lo, N. Balasubramanian, and D.-L. Kwong, "Si, SiGe Nanowire devices by top-down technology and their applications", IEEE Transaction on Electron Devices, vol. 55, pp. 3107, Nov. 2008.

22. Y. Tian, R. Huang, Y. Wang, J. Zhuge, R. Wang, J. Liu, X. Zhang, and Y. Wang, "New Self-Aligned Silicon Nanowire Transistors on Bulk Substrate Fabricated by Epi-Free Compatible CMOS Technology: Process Integration, Experimental Characterization of Carrier Transport and Low Frequency noise", International Electron Device Meeting, pp.895, Dec. 2007.

23. J. S. Becker, S. Suh, R. G. Gordon, "Highly conformal thin films of tungsten nitride prepared by atomic layer deposition from a novel precursor", Chemistry of materials, vol. 15, pp. 2969-2976, 2003.

ECS Transactions, 45 (4) 231-239 (2012)
10.1149/1.3700472 ©The Electrochemical Society

Non-Destructive, Large-Scale Imaging of Anti-Phase Disorder in GaP Epilayers on Si(001) Using Low-Energy Electron Microscopy

B. Borkenhagen[a], H. Döscher[b], T. Hannappel[b], G. Lilienkamp[a], and W. Daum[a]

[a] Institute of Energy Research and Physical Technologies,
Clausthal University of Technology, D-38678 Clausthal-Zellerfeld, Germany
[b] Helmholtz-Zentrum Berlin, Materials for Photovoltaics, D-14109 Berlin, Germany

We introduce low-energy electron microcopy (LEEM) as a surface-sensitive technique for the investigation of anti-phase disorder and other defects in III-V semiconductor layers on Si(100) and present first results for 40 nm GaP films on Si(100). Using an actively pumped ultra-high vacuum transfer chamber, we were able to preserve the surface structure and chemical composition of the MOVPE-grown GaP films on Si(100) during transfer from the MOVPE reaction chamber to the remote electron microscope. Applying dark- and bright-field imaging modes we demonstrate the unique potential of LEEM for non-invasive, fast and large-scale inspection of anti-phase disorder and surface defects in heteroepitaxial III-V systems on Si.

Introduction

The growth of high-quality III-V semiconductor layers on Si(100) substrates is a key technology for future optoelectronics and opens new perspectives for high-efficiency photovoltaics. GaP with an indirect band gap of 2.3 eV has strong potentials for lattice-matched tandem solar cells based on Si as substrate and active absorber material, as the band gap and the lattice constant of the III-V epilayer can be adjusted by addition of nitrogen in quaternary GaP(AsN) systems (1). For Si-based optoelectronic devices, binary GaP buffer layers on Si allow to accommodate subsequent optically active layers (2,3). With small concentrations of nitrogen, GaP-based III-V ternary (GaPN) or quaternary (GaP(AsN)) layers with a direct band gap can be grown on Si(100) (4,5).

A main obstacle for the growth of high-quality III-V layers on Si (100) substrates is the formation of anti-phase domains (ADP) in the III-V crystal as a result of the C_{2v} symmetry and the inevitable presence of steps on a Si(100) surface (6): monatomic Si steps separate orthogonally oriented but otherwise structurally equivalent adjacent terraces of the Si(100) surfaces and induce the growth of incongruent anti-phase domains in the polar III-V crystal (Fig. 1). APDs are separated by anti-phase boundaries (APBs) which consist of energetically unfavorable III-III and V-V bonds (Fig. 1) and typically propagate in growth direction.

Strategies to minimize anti-phase disorder have so far focussed on preconditioning of the Si substrate, mainly by using specifically misoriented Si(100) surfaces with double steps and a single-domain surface orientation which suppresses the formation of APDs to a large extent (7,8). Recently, Volz et al. reported the growth of GaP layers on well-oriented Si(100): in a multi-step process which involved the growth of a high-quality Si buffer layer they were able suppress anti-phase disorder in the GaP film when the film exceeded a thickness of 40-50 nm by burying the APDs in deeper layers of the film (9) but the influence of buried APDs on the performance of III-V devices such as solar cells remains unclear.

There exist a number of studies of anti-phase disorder in heteroepitaxial III-V layers on group-IV semiconductors. APDs and APBs have been imaged with high-resolution transmission electron microscopy (TEM) (10,11,12) and atomic force microscopy (AFM) (12,13). Statistical information about the spatial distribution was obtained from X-ray diffraction (14) while the average APD concentration in MOVPE-grown GaP films on Si(100) was determined by reflectance anisotropy spectroscopy (RAS) experiments performed *in situ* under MOVPE-conditions (8,15).

Recently, we have shown that low-energy electron microscopy (LEEM) is a promising method for imaging of APDs in MOVPE-grown films GaP films on Si(100) (16). Other than TEM and AFM, LEEM is non-invasive. It does not require elaborate sample preparation, and its lateral resolution is sufficiently high to yield microscopic informations about anti-phase disorder on macroscopic length scales. Here, we will present and discuss LEEM images obtained from 40 nm GaP films on Si(100) to demonstrate the potential of LEEM for fast, large-scale characterization of anti-phase disorder and surface defects in heteroepitaxial III-V/Si systems.

Experimental details

(100)-oriented GaP films of 40 nm thickness with a P-rich surface were prepared on Si(100) in a MOVPE reactor (AIX 200) with a custom-designed rapid sample transfer to a mobile UHV system for further transfer either to a separate UHV chamber with low-energy electron diffraction (LEED) and scanning tunneling microscopy facilities, or to our LEEM facility. As substrates we used nearly flat Si(100) wafers with a small misorientation of $0.1°$ towards the [011] direction. The small miscut suppresses formation of double-layer surface steps to a large extent. Details of the deoxidation process of the oxidized Si substrates under MOVPE conditions in hydrogen atmosphere and the preparation of a P-rich surface are described in Ref. (15). The average APD concentration of the GaP epilayer was determined in situ by reflection-absorption spectroscopy. After preparation and characterization, the heterostructure was transferred to a mobile UHV chamber (17). This mobile UHV chamber is equipped with a battery-driven ion getter pump to maintain a vacuum better than $2x10^{-9}$ mbar and allows a sample transfer from the MOVPE reactor to our remote LEEM facility without significant surface contamination as confirmed by unchanged diffraction patterns.

The low-energy electron microscope (LEEM) is conceptually similar to a TEM, but operates in reflection geometry and with much lower electron energies at the surface (typically 0.1 eV to 30 eV) to gain maximum surface sensitivity (18). The instrument can easily be switched between real-space surface imaging and μLEED imaging of the same illuminated area (18). For dark-field imaging, individual diffraction spots can be selected by an aperture in an intermediate diffraction image plane (19). The LEEM images discussed in this paper were obtained after mild annealing of the sample to remove weakly adsorbed contaminants.

Results and discussion

The low-energy electron diffraction pattern of a 40 nm GaP film on a slightly misoriented Si(100) substrate shown in Fig. 2 was observed with a conventional LEED optics after transfer of the heterostructure from the MOVPE reactor to a UHV system without breaking vakuum. The streaky diffraction pattern with (1/2,0) and (0,1/2) superstructure spots is characteristic of a two-domain, p(2x2)/c(4x2)-reconstructed, P-rich GaP(100) surface (20). Corresponding diffraction patterns were observed with the low-energy electron microscope (16) directly after transfer of the sample into the LEEM chamber. The (1/2,0) and (0,1/2) spots, respectively, of the two orthogonal reconstruction domains of the GaP(100) surface were used for dark-field imaging of anti-phase disorder of the epilayer with the LEEM.

Fig. 3 displays a dark-field LEEM image showing the lateral distribution of anti-phase domains. The striped pattern of main phase and anti-phase domains in GaP reflects the step-terrace structure of the slightly misoriented Si(100) substrate. The monatomic steps run parallel to the [01-1] direction and separate adjacent Si surface terraces of orthogonal orientation (Fig. 1). The average domain periodicity of the GaP surface, as determined from Fig. 3 and from corresponding APD structures in other regions of the surface, amounts to about 70 nm and agrees with the average terrace width of a Si(100) surface misoriented by 0.1° towards the [011] direction. Such a correlation between the distribution of APDs in the GaP epilayer and the domain-determining step structure of the substrate is possible if the anti-phase boundaries grow perpendicularly to the surface (21). Inspection of Fig. 3 also reveals that the domains of the anti-phase never exceed the average terrace width within a regular striped pattern as expected for monotamic steps of a properly miscut Si substrate with the misorientation specified above, while the main phase can locally extend over several terraces due to the specific step energetics of the Si(100) surface (22). The lateral distributions of anti-phase domains as observed in the LEEM images in Fig. 3 is very similar to plane-view TEM images of APDs of the same sample which have been compared in a previous publication (16).

Much different structures of anti-phase disorder were observed in dark-field images taken at different regions of the same sample with a larger field of view of 11-13 µm. Representative images of such APD structures are shown in Figs. 4 and 5. While the contrast of most parts of the imaged surface can be related to anti-phase disorder in the GaP layer, an almost regular arrangement of stripe-like APDs as shown in Fig. 3 with a width corresponding to the terrace width of the substrate is observed only for parts of the imaged area in Figs. 4 and 5. The most prominent appearance of APDs is that of target-like patterns of main phase and anti-phase domains in Figs. 4 and 5. In the neighborhood of target-like APD patterns the distribution of APDs is also changed, and main phase domains extend over much larger regions of up to several microns. Such large domains could be the result of extended substrate regions with double-layer step structure and, hence, terraces of unique domain orientation (8). A preferential formation of double-layer steps in a hydrogen ambient has been observed on Si(100) samples with a larger miscut towards the [011] direction (23). Presumably, the target patterns originate from GaP growth on regions of the Si substrate with almost perfect (100) orientation, where locally the step structure of the slightly misoriented surface does not follow the average step density given by the misorientation.

In addition to APDs which are responsible for most of the contrast in our dark-field LEEM images of GaP/Si(100), we observed mostly circularly or elliptically shaped defective regions of the GaP surface which showed a very different and non-complementary contrast when using (1/2,0) and (0,1/2) superstructure spots for dark-field imaging. A respective region can be seen in Fig. 4 close to the center of a target pattern. Such defects are even more prominent in bright-field images of the surface. While contrast due to anti-phase disorder is barely visible in the bright-field image in Fig. 5 (a), five to six prominent bright features with a dark core attributable to surface defects distinctly different from anti-phase disorder can be clearly seen. An assignment other than to APDs is also consistent with the dark-field images in Fig. 5 (b) to (e). In particular, for a primary energy of 10.2 eV the contrast of these regions is not reverted when imaging the surface with the (0,1/2) diffraction spot (Fig. 5 (e)) instead of imaging with the (1,1/2) diffraction spot (Fig. 5 (d)). This is strong indication for a different contrast mechanism in LEEM, and the surface electric field at and around the defects may play a much stronger role for imaging of these defects than for imaging of APDs.

In this context, the previous observations of the formation of liquid and metallic Ga droplets by other groups on both GaP (24) and GaAs (25) surfaces with LEEM is noteworthy. Tersoff et al. produced liquid Ga droplets on GaAs by annealing at 640°C and excess evaporation of As in ultra-high vacuum (25). In their LEEM investigations these droplets appeared with diameters between about 2 and 5 μm and with a bright ring, similar to the objects in Fig. 5 (a). Droplet formation of excess metallic Ga on our GaP surfaces appears to be a conceivable explanation for the round defects in Figs. 4 and 5. Such metallic precipitations alter the surface electric field and strongly affect the reflection of low-energy primary electrons, particularly in mirror electron microscopy (MEM) (18). It should be noted, however, that our heterostructures were prepared with an excess of P precursor gas to obtain a P-terminated surface, and that we observed the defects already prior to subsequent annealing in the LEEM chamber. It is unclear whether Ga droplet formation is possible under such conditions. In future work we will identify the chemical compositions of these defects by high-resolution scanning Auger microscopy. Future LEEM experiments will also address the question whether more of these defects are formed by heating the samples under UHV conditions in the LEEM chamber at moderate temperatures and whether the defects can be moved at elevated temperatures, similar to the Ga droplets in Refs. (24) and (25).

In a previous publication we have addressed similarities and differences between TEM and LEEM for imaging anti-phase domains in GaP/Si(100) and related III-V/Si systems (16). The resolution of TEM is superior to LEEM, although the resolution of present-generation low-energy electron microscopes has dramatically improved in comparison to the microscope used in this study. On the other hand, TEM requires an elaborate sample preparation, is destructive and provides information only on selected parts of the sample. In contrast, representative characterizations of APD distributions and other defects over large sample areas are comparatively easily performed with LEEM. For example, we have monitored the lateral distribution of APDs and the round defects in Fig. 5 (a) over a dimension of 200 μm by continuous translation of the sample using the bright-field mode. The possibility of fast imaging of large surface areas allows one to obtain statistically relevant information about the spatial distribution of APDs as well as about other defects and their distribution over the surface. Spatially resolved informations on APD distributions are important for further development and quantitative calibration of methods such as reflectance anisotropy pectroscopy which allows an *in situ* characterization of APDs during MOVPE (15) but yield signals averaged over a macroscopic scale.

Conclusion

We have applied low-energy electron microscopy as a non-invasive probe for large-scale characterization of anti-phase disorder in 40 nm GaP films grown by MOVPE on Si(100). Using a dark-field imaging mode, the spatial distribution of anti-phase domains was imaged. Besides more or less regularly striped APD pattern reflecting the monolayer step structure of the substrate, we observed target-like APD patterns and extended areas of single domain orientation, the latter presumably induced by substrate regions with double steps. In addition, round defects particularly prominent in bright-field images were found to be distributed over the surface. Future research will address the physical and chemical identity of these defects as well as strategies for their suppression. Moreover, LEEM investigations of GaP/Si(100) systems with different film thicknesses will enable us to characterize, on a large lateral scale, the evolution of anti-phase boundaries with film thickness.

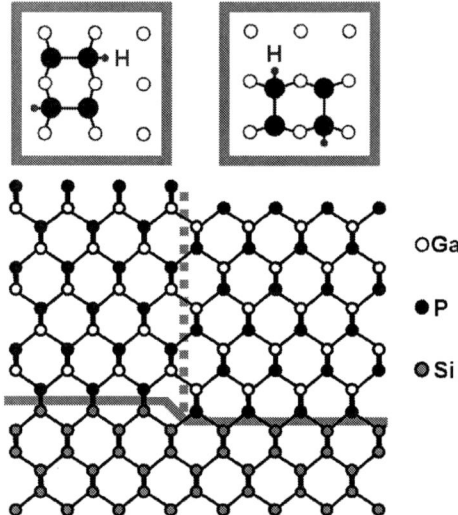

Figure 1. Lower part: side view of an anti-phase boundary in a GaP epilayer on Si(100) induced by a step of the substrate with single-atom height. As is the case for our samples, the GaP layer is P-terminated and the free valence of the P surface atoms saturated with hydrogen. The upper part represents top views of mutually orthogonal surface reconstructions of adjacent domains.

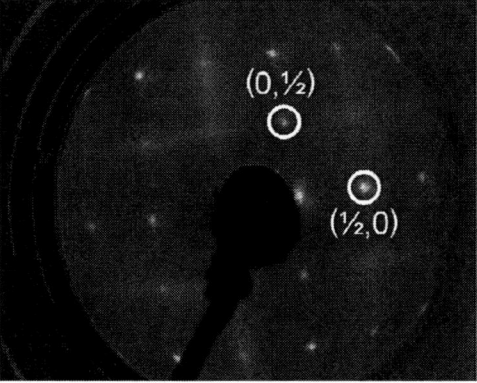

Figure 2. Low-energy electron diffraction (LEED) pattern of the reconstructed, P-rich surface of a two-domain GaP(100) epilayer on Si(100) obtained after CVD deposition with an electron energy of 60 eV. The (1/2,0) and (0,1/2) diffraction spots of the two orthogonal domains with mixed p(2x2)/c(4x2) surface reconstructions are encircled.

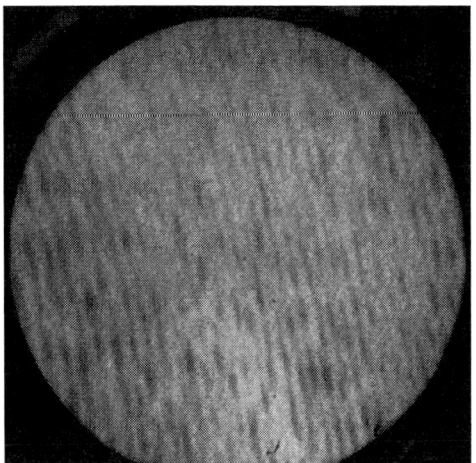

Figure 3. Dark-field LEEM images of 40 nm GaP/Si(100). Bright and dark regions represent the main phase and the anti-phase, respectively, showing anti-phase disorder induced by a regular array of single steps of the Si substrate. The field of view is 3 μm, the primary energy E_p of the electrons 67 eV.

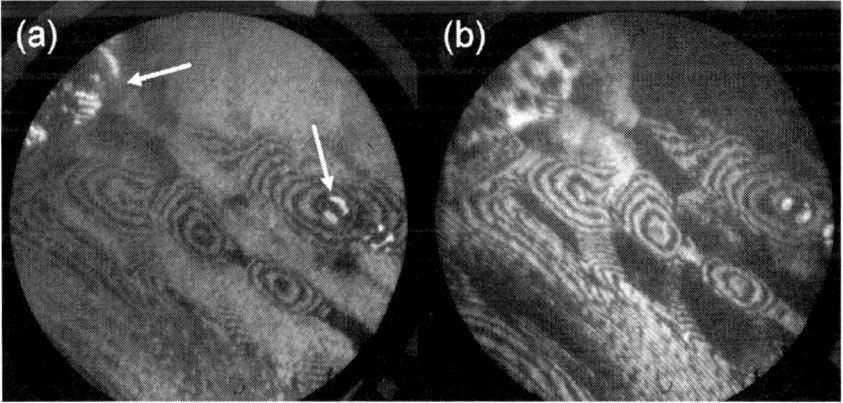

Figure 4. Dark-field LEEM images of GaP/Si(100) obtained with (1/2,0) and (0,1/2) diffraction spots, respectively, showing irregular and target pattern-like arrangements of main-phase and anti-phase domains. Besides anti-phase disorder, defects of different origin cause a different contrast with brighter intensity (arrows). The field of view is 13 μm, the electron energy 7 eV.

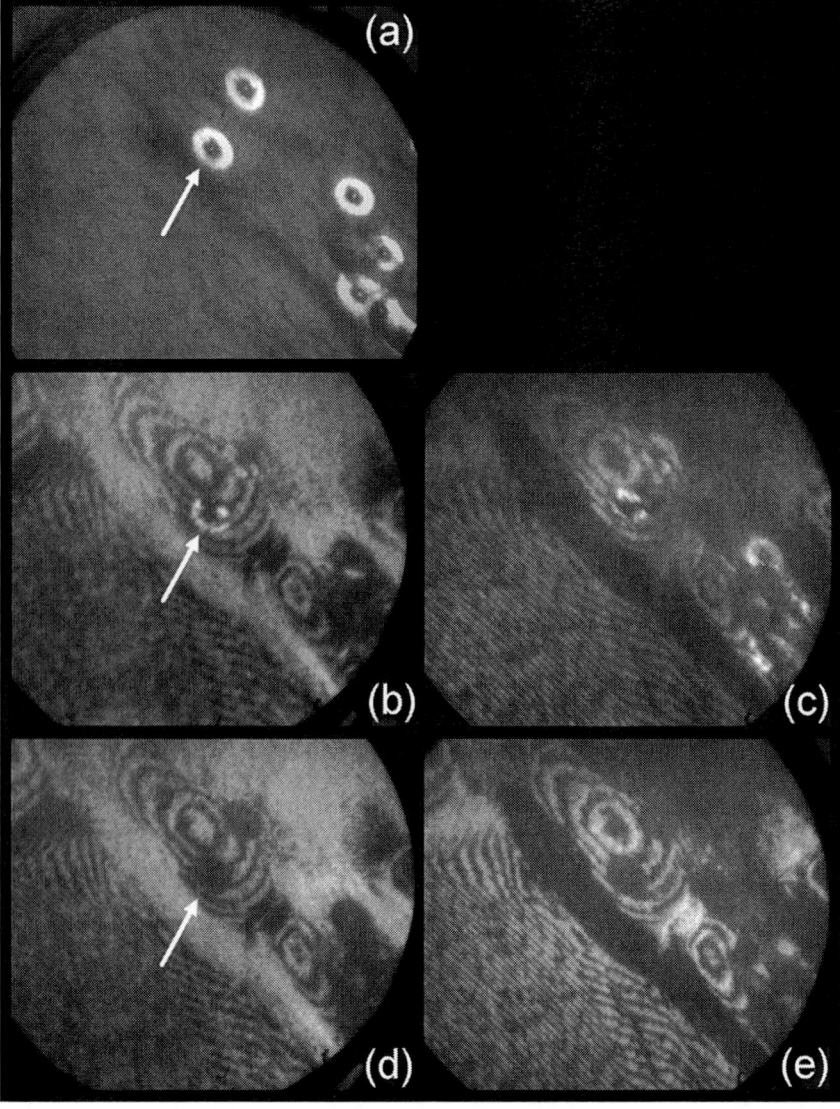

Figure 5. LEEM images of defect regions of GaP on Si(100) different from anti-phase domains. The field of view is 11 μm for all images. (a): bright-field image with $E_p = 4.35$ eV, (b) and (c): dark-field images with $E_p = 6.8$ eV, (d) and (e): dark-field images with $E_p = 10.2$ eV. As in Fig. 4, the (1/2,0) diffraction spot was used for the dark-field images on the left side, the (0,1/2) diffraction spot for the dark-field images on the right side. Arrows mark the same surface defect in bright- and dark-field images.

Acknowledgments

This work was supported by BMBF (T. H., Project No. 03SF0329C).

References

1. J. Geisz, J. Olson, D. Friedman, S. Kurtz, W. McMahon, M. Romero, R. Reedy, K. Jones, A. Norman, A. Duda, A. Kibbler, C. Kramer and M. Young, Conference Paper NREL/CP-520-36991, Jan. 2005.
2. Y. Takagi, H. Yonezu, T. Kawai, K. Hayashida, K. Samonji, N. Oshima and K. Pak, *J. Cryst. Growth* **150**, 677 (1995).
3. Y. B. Bolkhovityanov and O. P. Pchelyakov, *Physics-Uspekhi* **51**, 437 (2008).
4. J. Chamings, S. Ahmed, S.J. Sweeney, V.A. Odnoblyudov and C.W. Tu, *Appl. Phys. Lett.* **92**, 021101 (2008).
5. S. Liebich, M. Zimprich, A. Beyer, C. Lange, D.J. Franzbach, S. Chatterjee, N. Hossain, S.J. Sweeney, K. Volz, B. Kunert and W. Stolz, Appl. Phys. Lett. **99**, 071109 (2011).
6. H. Kroemer, *J. Cryst. Growth* **81**, 193 (1987).
7. J. Nemeth, B. Kunert, S. Reinhard, W. Stolz and K. Volz, *Thin Solid Films* **517**, 140 (2008).
8. H. Döscher, T. Hannappel, B. Kunert, A. Beyer, K. Volz and W. Stolz, *Appl. Phys. Lett.* **93**, 172110 (2008).
9. K. Volz, A. Beyer, W. Witte, J. Ohlmann, I. Nemeth, B. Kunert and W. Stolz, *J. Cryst. Growth* **315**, 37 (2011).
10. J. P. Gowers, *Appl. Phys. Mater. Sci. Process.* **34**, 231 (1984).
11. J. M. Zhou, H. Chen, F. H. Li, S. Liu, X. B. Mei and Y. Huang, *Vacuum* **43**, 1055 (1992).
12. I. Nemeth, B. Kunert, W. Stolz and K. Volz, *J. Cryst. Growth* **310**, 4763 (2008).
13. P. N. Uppal and H. Kroemer, *J. Appl. Phys.* **58**, 2195 (1985).
14. S. F. Fang, K. Adomi, S. Iyer, H. Morkoc, H. Zabel, C. Choi and N. Otsuka, *J. Appl. Phys.* **68**, R31 (1990).
15. H. Döscher and T. Hannappel, *J. Appl. Phys.* **107**, 123523 (2010).
16. H. Döscher, B. Borkenhagen, G. Lilienkamp, W. Daum and T. Hannappel, Surf. Sci. **605**, L38 (2011).
17. T. Hannappel, S. Visbeck, L. Töben, and F. Willig, *Rev. Sci. Instrum.* **75**, 1297 (2004).
18. E. Bauer, *Rep. Prog. Phys.* **57**, 895 (1994).
19. W. Telieps, *Appl. Phys. Mater. Sci. Process.* **44** 55 (1987).
20. L. Töben, T. Hannappel, K. Moller, H. J. Crawack, C. Pettenkofer and F. Willig, *Surf. Sci.* **494**, L755 (2001).
21. H. Döscher, O. Supplie, S. Brückner, T. Hannappel, A. Beyer, J. Ohlmann and K. Volz, *J. Cryst. Growth* **315**, 16 (2011).
22. B. S. Swartzentruber, Y.W. Mo, M.B. Webb and M. G. Lagally, *J. Vac. Sci. Technol. A* **7**, 2901(1989)
23. H. Döscher, P. Kleinschmidt and T. Hannappel, *Appl. Surf. Sci.* **257**, 574 (2010).
24. E. Hilner, A.A. Zakharov, K. Schulte, P. Kratzer, J.N. Andersen, E. Lundgren and A. Mikkelsen, *Nano Lett.* **9**, 2710 (2009).
25. J. Tersoff, D. E. Jesson and W. X. Tang, *Science* **324**, 236 (2009).

Author Index

Adley, D.	175	Eneman, G.	115
Ahn, J.	147		
Armstrong, B.	175	Floresca, H.	79
		Fuse, T.	203
Bain, M. F.	175		
Baine, P. T.	175	Gaddam, S.	49
Banerjee, S. K.	3	Geer, R. E.	31
Barnes, J.	165	Gergaud, P.	165
Baumvol, I.	137	Ghosh, S.	111
Bennett, B. R.	91	Ghyselen, B.	165
Birinci, E.	23	Gong, X.	209
Bogaerts, A.	73	Gu, J. J.	217
Bogumilowicz, Y.	165	Guo, C.	209
Boos, J.	91	Guo, H.	209
Borkenhagen, B.	231		
Brammertz, G.	115	Habib, K.	15
		Han, S. M.	111, 153
Carron, V.	165	Hannappel, T.	231
Caymax, M.	97, 115	Hassibi, A.	3
Chan, J.	79	Hellings, G.	115
Chen, D.	83	Heyns, M.	157
Colombo, L.	79	Hinojos, D.	79
Comfort, E.	63	Horiguchi, N.	97, 115
Cosemans, S.	157	Huyghebaert, C.	157
da Silva, S.	137	Jandhyala, S.	39
Datta, S.	129	Jouanneau, T.	165
Daum, W.	231	Jung, W.	3
De-Gendt, S.	97, 157		
Delabie, A.	97	Kelber, J. A.	49
Delaye, V.	165	Kim, J.	39, 79
Detavernier, C.	157	Kim, M. J.	79
Diebold, A.	63	Kittl, J. A.	157
Dimroth, F.	165	Klinger, V.	165
Döscher, H.	231	Kong, L.	49
Dougherty, D. B.	63	Krug, C.	137
Dowben, P. A.	49	Kumar, A.	91
Duffy, R.	189	Kuo, M.	129

Lake, R. K.	15	Reddy, D.	3
Lee, B.	39	Register, L. F.	3
Lee, J.	63	Riedinger, B.	23
Leonhardt, D.	111, 153	Rolim, G.	137
Li, J.	129	Rowe, J. E.	63
Lilienkamp, G.	231		
Lin, D.	97	Sandin, A.	63
Liu, H.	129	Saraswat, K. C.	91, 203
Liu, S.	83	Schaekers, M.	157
Lu, N.	79	Schwalke, U.	23
		Shayesteh, M.	189
MacDonald, A. H.	3	Shi, X.	157
Martens, K. M.	157	Shin, B.	147
Mayer, T. S.	129	Sioncke, S.	97
McIntyre, P.	147	Soares, G.	137
McNeill, D. W.	175	Sodemann, I.	3
Merckling, C.	115	Struyf, H.	97
Mertens, S.	157		
Meuris, M.	115	Tauzin, A.	165
Mitchell, S.	175	Thean, A.	97, 115
Mohata, D.	129		
Mordi, G.	39	Vallett, A.	129
Mou, X.	3	Verbruggen, J.	157
Nainani, A.	91	Waldron, N.	115
Nam, J.	203	Wallace, R.	79
Nelson, F.	63	Wang, G.	115
Neyts, E. C.	73	Wang, J.	79
Nguyen, N. D.	115	Wang, Y.	31
Nishi, Y.	203	Wessely, F.	23
Nyns, L.	97	Wessely, P.	23
		Winderickx, G.	115
Ong, P.	115		
Orzali, T.	115	Xue, Z.	83
Pasquale, F. L.	49	Ye, P. D.	217
Perova, T. S.	175	Yeo, Y.	209
Pesin, D.	3	Yuan, Z.	91
Prasad Sinha, D.	63		
		Zainal, N.	175
Radtke, C.	137	Zhang, M.	83
Radu, I. P.	157	Zhang, X.	209
Rampelberg, G.	157	Zhou, M.	49